Low Power Hardware Synthesis from Concurrent Action-Oriented Specifications

Gaurav Singh · Sandeep K. Shukla

Low Power Hardware Synthesis from Concurrent Action-Oriented Specifications

 Springer

Gaurav Singh
Intel Corporation
S. Mopac Expressway 1501
78746 Austin
TX, USA
gasingh@vt.edu

Sandeep K. Shukla
Virginia Tech
Bradley Department of Electrical &
    Computer Engineering
Whittemore Hall 302
24061 Blacksburg
VA, USA
shukla@vt.edu

ISBN 978-1-4899-8702-0   ISBN 978-1-4419-6481-6 (eBook)
DOI 10.1007/978-1-4419-6481-6
Springer New York Dordrecht Heidelberg London

Printed on acid-free paper

Springer is part of Springer Science+Business Media (www.springer.com)

*To the friends and families,
and all our fellow Hokies.*

Preface

Human lives are getting increasingly entangled with technology, especially computing and electronics. At each step we take, especially in a developing world, we are dependent on various gadgets such as cell phones, handheld PDAs, netbooks, medical prosthetic devices, and medical measurement devices (e.g., blood pressure monitors, glucometers). Two important design constraints for such consumer electronics are their form factor and battery life. This translates to the requirements of reduction in the die area and reduced power consumption for the semiconductor chips that go inside these gadgets. Performance is also important, as increasingly sophisticated applications run on these devices, and many of them require fast response time.

The form factor of such electronics goods depends not only on the overall area of the chips inside them but also on the packaging, which depends on thermal characteristics. Thermal characteristics in turn depend on peak power signature of the chips. As a result, while the overall energy usage reduction increases battery life, peak power reduction influences the form factor.

One more important aspect of these electronic equipments is that every 6 months or so, a newer feature needs to be added to keep ahead of the market competition, and hence new designs have to be completed with these new features, better form factor, battery life, and performance every few months. This extreme pressure on the time to market is another force that drives the innovations in design automation of semiconductor chips.

If one considers high-end servers, workstations, and other performance-hungry systems, their corresponding semiconductor chips also need to resolve many of these market and technological forces. However, the time scale and the scale of betterment from generation to generation are quite different. In this book, our focus is not on these high-performance computing systems. Our focus is rather on the specific, battery-driven consumer devices, in particular the semiconductor ICs or chips inside them.

However, we are not concerned here with the architecture or the design specifics of these chips, but rather with the process of designing them. These are mostly implemented as ASICs (Application-Specific Integrated Circuits) or sometimes on FPGAs (Field Programmable Logic Arrays) for computing specific functions or algorithms. For example, one could design a chip for encryption or decryption of bit streams or one could design a chip for matrix multiplication or Fast Fourier

Transform. One could also have complex interfaces in these chips, such as an AMBA bus interface, or a DMA interface, or some other kind of communication or I/O interface. These would mean that not only such chips do specific computation, they also have multiple threads of control. The computation itself may also be designed with multiple threads of control, because hardware does allow us to overlap computations that do not have to be sequenced in a specific order. However, such concurrent threads of computation require synchronization and communication with each other. Most computation and communication functions have such concurrency and synchronization between concurrent threads of computation.

The question is how to design such a highly concurrent hardware system and fabricate it on a chip, while resolving various constraints on performance (e.g., latency), power consumption (e.g., peak power, leakage, overall energy), area, etc. This is a question of innovative design methodology, design environment, language for design entry, transformation, abstraction, and many other issues. Electronic Design Automation (EDA) community has been debating these questions for many years, but in the last 10 years or so, this question has become one of the prime movers of the industry. Designing complex chips that satisfy requirements and design constraints and delivering them fast enough to meet the time-to-market goals is the central issue of design automation these days.

Semiconductor IC design process has gone through a long history in the last 40 or so years, starting from SSI (Small-Scale Integration), MSI (Medium-Scale Integration), LSI (Large-Scale Integration), VLSI (Very Large-Scale Integration), ULSI (Ultra Large-Scale Integration), etc. Moore's law as predicted by Gordon Moore is still going on with full steam, and we are seeing a doubling of the number of transistors per unit area of the chip every 2 years approximately. As a result, more complex functionalities are being implemented on chip, and thus the need for ultra large-scale integration – UVLSI. (The term UVLSI did not catch up, and the term VLSI is still pervasive in the literature, even though since the time the term VLSI was introduced, the scale has improved by orders of magnitude.)

In the beginning, the designers used to design each transistor by hand, tweaking their parameters to make sure that required characteristics are implemented correctly. Then came gate-level design era, where schematics were initially hand drawn, and later various automated tools arrived. However, as the industry moved from hundreds of gates to thousands of gates on the die, the gate-level design techniques and corresponding automation processes also ran out of steam. The advent of register transfer level (RTL) languages in the early 1990s ushered a new era of productivity in silicon chip design. Capturing the state of the system with registers and state transition conditions with combinational logic facilitated the designers' ability to make much faster design entry into the automation flow. Initially, this abstraction from gate-level design to register transfer level was meant to provide better simulation abilities. A design described only with gates would take much more time to simulate for a few million clock cycles than if it were described with RTL. VHDL was conceived as a simulation language, and so was Verilog. Pretty soon various logic synthesis algorithms started to show up in the literature, and eventually logic synthesis allowed the large-scale proliferation of the use of RTL as a design entry language for most semiconductor IC companies by the late 1990s. The use of RTL thus reduced validation time by speeding up simulation and reduced the design time by virtue of optimized logic synthesis algorithms and tools.

However, the designers started to demand more out of the automation tools and tried to introduce higher level constructs into RTL, and *behavioral RTL* constructs found ample designer attention. The original variant of RTL called the *structural RTL* described the architecture of the design in terms of modules, the combinational logics with interconnected gates, and the state machines in terms of registers and their update logic. Behavioral RTL aimed at easing the burden on the designers by allowing them to describe the designs in terms of their behaviors rather than the architectural structures (e.g., a state machine described as a state transition system, and not as an interconnection of registers and gates). Algorithms for logic synthesis from behavioral RTL became a popular topic of research. Given a behavioral state machine description, the synthesis algorithm is required to choose among various possible structural implementations of the state machine. For example, one could vary the state encoding (e.g., binary vs. one-hot vs. Huffman encoded) and obtain various area, power, latency characteristics of the design. Instead of the designer having to make the choice, the synthesis algorithm is supposed to do a design space exploration and choose the best implementation based on the design constraints. This means that the synthesis algorithm had to be made aware of constraints such as area, power, latency. The exploration, however, is non-trivial because depending on such design constraints, one encoding vs. another would be more appropriate. This meant that the synthesis algorithms have to solve multi-objective optimization problems before making decisions on what to synthesize. A number of tools such as *Behavioral Compiler* attempted this, but did not gain popularity because expert designers thought that they could do a better job through their experience and intuition on what makes the best hardware for the given set of constraints.

One of the most important reasons for promoting the behavioral style of RTL came from the fact that the chips being designed got increasingly complex, requiring various parts executing sophisticated protocols for communicating with the outside world as well as among its own components. Protocols are best described behaviorally than structurally. Thus *protocol compilation* was another name for behavioral synthesis. There has been mixed reaction to such tools and methodologies, and it was suspected by expert designers to be producing non-optimal implementations.

In the late 1990s and early part of 2000, a push for higher abstraction level than RTL came about. A number of activities related to C/C++-based design entry languages for hardware were announced, including SpecC from the University of California – Irvine, Synopsys' Scenic, later named SystemC, Cynapp's Cynlib, IMEC's OCAPI, etc. Also, enhancing Verilog into Superlog and VHDL into Object-Oriented VHDL were announced around the same time. It was clear that without raising the abstraction level for hardware design entry, it is hard to keep up with the increasing productivity requirements in the semiconductor industry.

The introduction of high-level software languages as carriers for RTL description does not, however, enhance the abstraction level. It only allows one to compile the hardware description with traditional software compilers and, therefore, provides a free simulation environment. But the cost of RTL simulator is not the biggest concern for the industry, and hence a need for higher abstraction level was given serious thoughts. More academic languages such as SpecC already had introduced a number of conceptual abstractions such as behaviors as processes, channels for communication, channel refinement for communication protocol elaboration, events, and various synchronization between behaviors and events. A stepwise refinement strategy and taxonomy formed the core of the SpecC methodology. Some of these abstractions found their way into the SystemC-2.0 specification and ushered the era of transaction-oriented design entry. The terminology of transaction level model or TLM came about very soon after that, and various levels of transaction-level descriptions were prescribed and adopted into the SystemC language.

The transaction-oriented description of functionalities of a hardware system (or hardware/software system) allows one to abstract away from descripting the bit-level details of the design, especially the communications between modules within a design. Transactions not only allow abstracting away the data-type representation from bits to high-level data types but also allow temporal abstraction. For example, a data transfer over a bus between a module and memory could take a number of clock cycles, and the protocol may be described by a cycle-by-cycle description of what bits get set and what bits get reset at each cycle. This entire process can be replaced by a simple transaction. As a result, transaction-level models can be simulated much faster (orders of magnitude) compared to simulating RTL models.

Since transactions abstract away precise cycle-by-cycle behavior, synthesizing code from transactional model into real hardware brought the old pain back. Now, the synthesizer has to solve multi-objective optimization problem to select from various possible elaboration of transactions into bit level, cycle-by-cycle behavior. If such optimization problem is not solved fast and accurately, the synthesized hardware will be possibly suboptimal. This became a cause of pain for a while, and even today, this problem is not entirely solved. However, progress has been made in two directions. Synthesis of acceptable-quality RTL from TLM models has been done by a number of companies – Forte Design Systems' (now erstwhile Cynapps) Cynthesizer has much of such capabilities. Mentor's Catapult-C, initially started as an algorithmic C to RTL synthesizer, also has adapted itself so it can handle a lot of transactional constructs and synthesize quality RTL. The second technological innovation that allowed this trend of TLM-based design entry to proliferate further was Calypto System's sequential equivalence verification engine that can compare a TLM model and the generated RTL to verify their sequential equivalence.

Having reached this status has enabled the design industry, especially ASIC industry, to progress toward bridging the notorious productivity gap in the industry. However, a lot more is desirable, and TLM with higher abstraction is still out of the scope for much of the automated synthesis tools. TLM synthesis and verification and transference of the verification assurance at the TLM level to the resulting RTL level are still topics of research. Such research papers are in abundance at most design automation-related conferences today.

Recall that the ability to synthesize is not the only criteria for success here, nor is the ability to carry out automated sequential equivalence verification. The quality of the generated RTL in terms of latency, power, area, etc., is very important, and active research is being carried out along those lines. Many of the tools mentioned above also allow the users to specify timing, power, and area budget constraints and accordingly try to synthesize RTL that meets those bounds. However, estimating power, area, or timing from a TLM-level model is another hard problem. For example, power consumption of a hardware design is highly dependent on the technology being used, details of the physical structure of the design, clock frequency, voltage levels, interconnects and clock tree, etc. Therefore, optimizing the result of synthesis from TLM models for power has to be based on a number of assumptions, profiling of past designs, and various intuitions formulated as numeric guesses. Many of these have been formalized in terms of statistical regressions over other parameters such as toggle counts, state transitions, transaction duration, and width, and research in this field is being carried out vigorously as we write this preface.

As this push toward abstraction was warming up, an alternative approach started taking shape at the Massachusetts Institute of Technology (MIT). James Hoe and Arvind started looking at the specification of hardware in terms of *term rewriting systems*. Term rewriting is a topic extensively studied by the automated theorem-proving community. For example, the steps to proving an algebraic identity can be captured by rewriting rules. Starting from the left-hand side of the identity to be proven, one applies appropriate rewrite rules to arrive at the right-hand side term. Computer algorithms that efficiently figure out which rewrite rules to apply in what order are at the core of term rewriting systems. It seems that the intuition behind Hoe and Arvind's work was as follows. One could conceive of the current state of a hardware system as a term over an appropriate term algebra and the transitions of the system as rewrite rules. Therefore, the evolution of a hardware system in its state space can be looked upon as applications of rewrite rules from the term denoting the initial state. This approach to hardware specification turned out to be abstracting the system in a direction that transactions usually cannot. The concurrency in hardware systems (i.e., one set of registers changing state independent of another set of registers changing state, at the same clock cycle) gets captured by the different rewrite rules. One could actually apply multiple of these rules to the same term (state) to express concurrent evolution in the state space. More importantly, the parts of the state space (registers) on which the multiple rules are applied must be disjoint; otherwise, there will be attempts to change the state of the system by two distinct transitions at the same time, leading to *race conditions*. This *non-interference* property is essential for representing the correct transitions in the state space. In the concurrency literature this is known as *atomicity* of the transactions, or the rules. Finding which rules can be applied concurrently without violating atomicity is not difficult, if one knows exactly which parts of the state these rules try to modify. One can declare those rules which modify overlapping parts of the state space as "conflicting" and create a conflict graph. The conflict graph will have nodes denoting rules and edges denoting conflict. Thus, finding the rules which can be concurrently applied in the same clock cycle is equivalent to finding the *maximal independent set* in the conflict graph. Finding the maximum size-independent set is an NP-complete problem, but finding maximal such set can be heuristically done efficiently. This led to the scheduling algorithms required to synthesize RTL from such kind of hardware specification. Hoe and Arvind created a new paradigm of hardware specification and synthesis of RTL from such specifications by computing schedules of rules that apply in each clock cycle. The RTL they created had this scheduler inbuilt into the hardware. Later this concept evolved into the language *Bluespec*. The Bluespec language was first used in a network processor design company and was later developed into a product by an EDA company, also named Bluespec Inc. Both of these companies were started by Arvind and his colleagues.

It turns out that the idea germane to the term rewriting view of the state transitions had other variants in the literature. Dijkstra had introduced the notion of a guarded command language, where a system state is spread across multiple variables. A guarded command is a rule that updates those variables based on certain conditions. Given a set of guarded commands, the update process is based on rounds of non-deterministic choices. The choice at each round is among the set of guarded commands whose guard evaluates to true over the current state. This is a simple model to describe interleaving concurrent evolution of the state space. Even if one command is executed per round, it could model concurrent execution of commands. For two successive commands that do not overlap in the set of variables they modify, the effect of executing them sequentially is equivalent to that of executing them concurrently. Chandi and Mishra later extended this notion and defined the UNITY language based on guarded commands for parallel program specification. Any possible execution sequence of commands allowed by the program is a behavior of the specified program. An implemented program, however, may not have all these behaviors but some of the possible ones. Thus, an implementation satisfies such a specification if the set of possible behaviors of the implementation is a subset of the set of possible behaviors of the specification.

A number of computer scientists including Leslie Lamport suggested that interleaving semantics of parallel programs is the appropriate semantics. This means that if you have two guarded commands a and b which can execute at the same time in parallel, then the effect of that is equivalent to first executing a and then b or vice versa. Diagrammatically, this leads to a diamond representation, often called the diamond rule. If a and b are non-interfering (i.e., they do not modify the same variables or state elements), it makes sense to view their execution this way. However, if execution of a modifies a variable that is being used in the guard of b, then after execution of a, the guard of b may not hold true any more. Hence, it would not be possible to have the same effect as a and b executing in parallel. In the interleaving semantics, even parallel execution of a and b would not be allowed in such a case, because there is no equivalent interleaving of such guarded commands.

The other school of thought about parallel program specification at that time was championed by Vaugh Pratt at Stanford and his colleagues. This was termed "partial order" semantics of concurrency. According to this semantics, interleaving semantics misses out many possible interactions between concurrent executions and is not sufficient to capture all possible behaviors one could see in concurrent systems. Also, they came up with examples of scenarios where two concurrent systems that are indistinguishable in terms of interleaving semantics are distinguishable by partial order semantics, based on partially ordered multiset (POMSET) models of behaviors.

There was a raging debate between these two schools in the early 1990s over an Internet-based mailing list called the "concurrency mailing list." The often heated exchange of messages on the mailing list between Pratt and Lamport is interesting to read and is available in the appendix of a book on partial order models of concurrency edited by Holzmann and Peled, published in 1996. However, here we need not concern ourselves with POMSETs and partial order semantics. Hardware system behaviors are usually observed as traces of inputs and outputs as the state space evolves in reaction to the inputs that are provided to it by the environment. As a result, a trace-based semantics is sufficient for our purposes.

A lot of the work described in this book is based on our experience in working with the Bluespec engineers and researchers. However, in order to make the model of concurrent atomic rules (also called actions) independent of a specific company, we adopted the idea of CAOS (Concurrent Action-Oriented Specifications) which is a Bluespec-like guarded atomic action-based language. In this language, the actions have guards like guarded command languages and Bluespec. The guards are predicates evaluated over the state of the system. At every round of execution, guards are evaluated, and the rules or actions whose guards evaluate to true are called *enabled actions*. The execution of the rules can be carried out in many different ways. The simplest one is to choose one of the enabled rules non-deterministically, execute it, and record the change in the state. In the next round, guards are reevaluated and a rule is selected among enabled ones. This goes on ad infinitum or until no guard is evaluated to true in a round. This is considered the *reference semantics* of any CAOS model. This describes all possible allowable behaviors of the specification. If a certain behavior is undesirable, the actions and their guards may be suitably modified, and possibly more state elements are added to modify the model such that undesirable behaviors are eliminated.

If one's intention is to create a hardware system implementing the specification, it is not a good idea to execute one rule at a time. This is because one would map each round of the model execution to one clock cycle of the actual hardware. Therefore, executing one rule at every cycle will provide an unacceptable latency of the system. To obtain better latency, one should attempt to execute as many rules as possible per round. This reduces to determining which rules are non-conflicting and then selecting a maximal set of such rules per round (clock cycle). But one has to be careful at this point. We have declared the behaviors generated by executing one rule per round as the reference semantics of our model. Thus, by executing multiple non-conflicting rules in the same round, if we create a behavior which has no equivalent behavior in the reference semantics, we will be violating the specification. As a result one has to worry about an extra constraint while scheduling – even when two rules *a* and *b* are non-conflicting, one is allowed to execute them concurrently in the same round, if and only if the resulting change in state can also be effected by either executing *a* first, and then *b*, or vice versa. This is important for correctness.

Thus, the scheduling process must know which pair of non-conflicting rules would violate this particular constraint and never schedule them together in a single round. A simple example would be two rules written as follows: R_1 : true $\rightarrow$ $x := y$ | R_2 : true $\rightarrow$ $y := x$. These two rules R_1 and R_2 are non-conflicting and when executed concurrently, they swap the values of x and y. However, if you execute R_1 followed by R_2, then x and y both end up having the last value of y. If you do the opposite, both x and y end up having the last value of x. Hence, the rules R_1 and R_2, albeit non-conflicting, cannot be executed in the same round, concurrently.

Once this restriction is understood, creating a scheduler and embedding it into the generated RTL is not very difficult. Thus, synthesizing RTL from CAOS or Bluespec is not a problem at all. In fact, in a 2004 ICCAD paper, Arvind et al. showed that the generated RTL using Bluespec is often competitive in terms of area and latency against the handwritten Verilog RTL for the same functionality. More interestingly, the generated RTL's structure is quite easy to understand.

As mentioned before, the computation of the schedule that executes maximum possible number of rules in every cycle is computationally hard but the heuristics perform well in most cases. One could, however, show contrived examples where even the heuristics can produce results with latencies arbitrarily far from the optimal. But such contrived examples do not happen in practice, and the Bluespec synthesis does produce hardware that has required area and latency. One of the chapters in this book will look deeper into the algorithmic complexity of the scheduling problems related to such CAOS-based synthesis and their approximability with heuristics. Such analyses are important for the sake of understanding the heuristics and their corner cases, but in practice they do not make much difference.

The 2004 ICCAD paper claimed nothing about the power consumption or peak power of the designs generated using Bluespec. That is where this book contributes the most to the CAOS-based synthesis processes. Since executing all non-conflicting rules in the same cycle might reduce latency, there may be a tendency to do so. But that may have several implications – the peak power of the hardware will rise, leading to thermal issues, degradation of performance, and stronger cooling and packaging requirements. Thus, it is not inconceivable to slow down the system a bit by not scheduling all the rules that could be scheduled in the same cycle, but rather stagger them a bit. To do this, one has to have an estimate of the power consumption of each hardware resource that is engaged in the execution of a rule, create a metric of peak power per rule, and accordingly solve an optimization problem. More interestingly, selecting some rules for execution and leaving the rest for later cycles might mean that the behavior of the design has changed. Nevertheless, such selective execution of the rules in a clock cycle can be done as long as the resultant behavior is still included in the set of possible behaviors under the reference semantics.

This led us to another interesting work described in this book. How do you know that when you change the behavior of a design by disallowing some enabled rules from executing in a clock cycle, you are still adhering to the original specification? Neither Bluespec nor CAOS had a formal verification methodology at that point. In fact, the question of formal verification is a bit tricky in this context. One possibility is to formally verify the generated RTL using standard model checking techniques. But this does not provide any advantage over standard RTL verification techniques used in other methodologies for hardware design. Since the one rule per round semantics embodies the reference semantics for the specification, it would be best to prove that the set of behaviors produced by the synthesized hardware is a subset of the behaviors allowed by the reference semantics. This means that standard model checking is not what we want, but rather automata theoretic language inclusion-based formal verification is more appropriate. There are not many tools out there which (i) allow automata theoretic formal verification and (ii) have the facility to model concurrent atomic actions in a straight-forward manner. The one freely available tool which has a great reputation in software verification is SPIN from Bell Labs. However, SPIN does not have a notion of clock cycle, and hence requires addition of a clock process if clocked hardware needs to be modeled. Having managed these modeling issues, we were able to model the reference behavior and the behavior of the generated hardware into SPIN and carry out automata theoretic verification. The interesting results we obtained were as follows: if a maximal set of non-conflicting rules are scheduled per cycle then the set of behaviors we obtain is correct with respect to the reference model. So is the peak power saving-based scheduling model which selectively executes rules based on a peak power constraint. But the peak power saving model and the standard RTL model (which executes maximal set of non-conflicting rules) actually may not have the same set of behaviors. This is a bit of a conundrum, because one would think that peak power saving model would have behaviors that are special cases of the standard RTL behavior, but this is not the case always. This is due to the unique nature of the CAOS-type modeling paradigm. The model is non-deterministic in its reference semantics, and hence a large number of different behaviors are allowed, and any implementation could actually pick to implement a subset of these behaviors. So two distinct implementation strategies might produce very different behaviors, both of which are correct with respect to the reference specification.

One way out of this dilemma would be to always use the same synthesis strategy, so that behaviors of the synthesized RTL are always the same. But that means any new optimization trick would be difficult to incorporate into the synthesis process. Another way is to suitably restrain the CAOS model with lots of constraints in the guards, such that the model has deterministic behavior. In other words, the reference semantics has only a very restricted set of behaviors. But this means constraining the CAOS model very tightly and burdening the designer with all the worries of handling synchronizations to determinise the model. We suspect that such an approach would make hardware specification with CAOS as error prone and hard as with traditional RTL specification.

As we alluded to earlier, peak power reduction is an important aspect of chip design due to thermal and packaging considerations. But overall energy reduction for increasing battery life is also another aspect of power optimization. This field is well studied for over a decade, especially with clock-gating and operand isolation techniques. Clock-gating essentially helps in reducing register power, by gating the clocks to the registers that do not change their values in a particular clock cycle. The gating logic can be inserted automatically during the synthesis process. Operand isolation is a process of isolating combinational logic from its inputs when the corresponding output is not being used in the current clock cycle. Both of these techniques have shown to save average dynamic power. For CAOS-based synthesis, one could essentially implement these techniques by analyzing the operations involved in each action or rule. Specifically, when the guard of an action is evaluated to false in a clock cycle, we know that the corresponding state updates will not take effect in that cycle. However, if there are other rules which might also update the same state element (register) and if one of those rules is enabled and selected for execution by the scheduler, then one has to make sure that the clock-gating of that register does not get turned on in that particular cycle. This requires more involved analysis of data dependency between various rules. In order to take some of these decisions efficiently during the execution of a design, extra logic needs to be added, which in turn is a potential source of additional power consumption. Thus, one has to figure out if the additional circuitry added to enable clock-gating and/or operand isolation is not offsetting the gains in overall power savings for the design. These techniques were implemented in the Bluespec synthesis tool called *Bluespec Compiler*. With specific options enabled in the compiler, one could effectively reduce the power consumption of the generated design. Power estimation techniques applied on various industrial benchmark designs have shown power gains. This book has a chapter on this particular topic reporting the techniques and detailed experimental results.

To sum up, in our experience, every time a new paradigm of modeling is introduced for hardware design entry, one has to create a design flow starting at specification of designs in that paradigm. But a design flow is not possible without automated synthesis techniques to generate the implementation at gate level or RTL. Once the synthesis algorithms are proven to be correct, one has to concentrate on resource optimizing synthesis techniques, especially targeting reduction in area, latency, power, etc. One might have to consider various trade-off points and accordingly create the appropriate tools and methodologies. The work described in this book is on CAOS-based hardware specification and was inspired by our association with Bluespec language and the company. Therefore, this work is focussed on various power reduction issues related to CAOS-based synthesis and the corresponding formalism, algorithms, complexity analysis, and verification problems.

We believe that it will provide the readers, especially research students who are entering the field of resource-constrained synthesis of hardware from high-level specifications, with a perspective and guide them into creating their own research agenda in related fields.

We are particularly indebted to Bluespec Inc. for their financial support in carrying out this research work [105–111] between 2005 and 2007, including multiple summer internships for the first author. A part of this research was also supported by a National Science Foundation PECASE award and a CRCD grant. We also received personal attention and help from Arvind, Rishiyur Nikhil, Joe Stoy, Jacob Schwartz, and others at Bluespec Inc. Sumit Ahuja at the FERMAT lab of Virginia Tech has been particularly helpful by providing many suggestions in power estimation and power reduction techniques. Finally, this work was the basis of the Ph.D. dissertation [103] of the first author, and this Ph.D. work was carried out at Virginia Tech's Electrical and Computer Engineering Department. We are indebted to Virginia Tech for all the facilities.

Austin, TX   Gaurav Singh
Blacksburg, VA   Sandeep K. Shukla
March 2010

Acknowledgments

We acknowledge the support received from Bluespec Inc., NSF PECASE, and NSF-CRCD grants, which provided funding for the work reported in this book.

Contents

List of Figures

List of Tables

Acronyms

RTL	Register Transfer Level
HLS	High Level Synthesis
EDA	Electronic Design Automation
HDL	Hardware Description Language
CAOS	Concurrent Action Oriented Specifications
CDFG	Control Data-Flow Graph
HTG	Hierarchical Task Graph
GCD	Greatest Common Divisor
BSC	Bluespec Compiler
BSV	Bluespec System Verilog
LPM	Longest Prefix Match
VM	Vending Machine
LTL	Linear-time Temporal Logic
TLA	Temporal Logic of Actions
MCS	Maximal Concurrent Schedule
ACS	Alternative Concurrent Schedule
MNS	Maximum Non-conflicting Subset
MIS	Maximum Independent Set
MLS	Minimum Length Schedule
FFD	First Fit Decreasing
PTAS	Polynomial Time Approximation Scheme
AES	Advanced Encryption Standard
UC	Upsize Converter

Chapter 1
Introduction

1.1 Motivation

A typical hardware design flow starts with the description of a design at a particular level of abstraction which is then synthesized to the corresponding low-level implementation of the abstract description. For example, RTL (Register Transfer Level) descriptions of hardware designs are synthesized to gate-level implementations, which are further synthesized to the physical level. Thus, hardware designs are described at different levels of abstraction. The abstract description of a design offers the benefits of ignoring low-level details resulting in faster architectural exploration and simulation. On the other hand, low-level implementation presents a more detailed and accurate view of the design. Such a top-down approach is commonly used for the generation of most hardware designs.

In most cases, designs are initially modeled at RTL since such models can be successfully synthesized to downstream gate-level and physical-level models using appropriate synthesis tools. However, as the chip capacities increase in adherence to Moore's law, RTL is becoming too low level for designing complex hardware designs. Consequently, designers are forced to look for innovative methods to handle the complexity of hardware designs. One way of addressing the complexity issue is to raise the abstraction level of a hardware design model even further by enabling the synthesis of RTL models from the corresponding high-level behavioral specifications of the designs. High-level specifications allow the designer to ignore various RTL coding issues (such as synchronization and scheduling of operations) while writing the description of hardware designs. Such high-level models aid in easier and faster analysis of the functionality of complex designs. Due to efforts in this direction, the process of such high-level synthesis of hardware designs is gaining considerable traction both in industry and in academia.

At high level, decisions about various design constraints and trade-offs can be taken more efficiently, which aids in faster architectural exploration, thus reducing the overall design time. Various area, latency, and power optimizations can also be targeted during the synthesis of such high-level behavioral models. Moreover, verification of high-level models results in the removal of various errors early in the design cycle. This can be used to ensure the correctness of the high-level specifications of the design, thus avoiding the existence of bugs in its low-level

G. Singh, S.K. Shukla, *Low Power Hardware Synthesis from Concurrent Action-Oriented Specifications*, DOI 10.1007/978-1-4419-6481-6_1,
© Springer Science+Business Media, LLC 2010

implementations which are relatively harder to debug. Hence, use of high-level models for synthesis of hardware designs is a promising technique for handling complexity issues of hardware designs, thus increasing the designer's productivity.

A variety of high-level synthesis methodologies exist which includes synthesis from high-level C-like Specifications [45, 49] and Esterel Specifications [15]. Recently, synthesis from Concurrent Action-Oriented Specifications (CAOS) is proposed which offers the advantage of inherent modeling of a hardware in terms of concurrent actions. Synthesis from CAOS has been shown to produce RTL code comparable to handwritten Verilog in terms of area and latency [10]. However, power-optimized synthesis and verification related to such a synthesis process are other two important issues which need to be addressed for the success of CAOS-based high-level synthesis. As discussed in the rest of the chapter, this book primarily focuses on power optimization as well as associated verification issues related to high-level synthesis from CAOS.

1.2 High-Level Synthesis

High-level synthesis (e.g., [28, 78]) allows the designer to write high-level (above RTL) behavioral descriptions of a design that can be automatically converted into synthesizable RTL code [95]. High-level code describes the operation on different data types of a hardware design without specifying the required resources and schedule of cycles. Thus, high-level synthesis offers the advantage of automatic handling of the scheduling and synchronization issues of a design, which helps in faster architectural exploration leading to reduced design time and increased productivity. Figure 1.1 shows the typical high-level to gate-level design flow used in

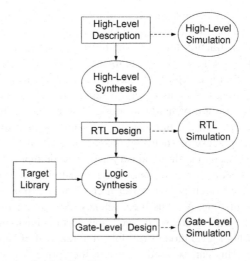

Fig. 1.1 High-level to gate-level design flow

the hardware industry to generate hardware designs. Gate-level designs are further synthesized using other downstream tools to generate ASICs or FPGAs.

Some examples of such high-level synthesis tools are Mentor's *Catapult* [82], *Bluespec Compiler* [18], Forte's *Cynthesizer* [52], *Esterel Studio* [50], Synfora's *PICO Express* [114], Celoxia's *Agility Compiler* [23], NEC's *Cyber* [117], and *SPARK* [57]. Depending on the synthesis tools, different languages are used to describe the high-level specifications of the hardware designs which are then passed as inputs to these tools. In some cases, extensions of traditional Hardware Description Languages (HDLs) such as Verilog or VHDL are used to efficiently describe the behavior of the designs at a high level of abstraction. For example, System Verilog is one such extension of Verilog. In addition to all the features of Verilog, it supports the concepts of object-oriented programming (including classes and interfaces), assertions, etc. Such languages are targeted for easier high-level modeling by hardware designers accustomed to using traditional HDLs. On the other hand, many designers often write system-level models (which can contain both hardware and software components) using programming languages, such as C or C++, for verifying the performance and functional correctness of the system. In order to accelerate the design process, automatic synthesis of hardware components from such system-level models is preferred. For this, various synthesizable subsets and/or extensions of C and C++ such as *SystemC* [56], *SpecC* [53, 88], *HardwareC* [74], *Handle-C* [24] are used to describe the behavior of the design.

A typical high-level synthesis process starts by generating the intermediate representations in the form of various flow graphs such as Control Data Flow Graphs (CDFGs), Hierarchical Task Graphs (HTGs) (which are hierarchical variants of CDFGs) in order to capture the data dependencies and control flow in the high-level specifications of the design given as the input. Various optimizations and transformations are then performed on such flow graphs before generating the RTL code. In this book, we call such synthesis techniques as CDFG-based high-level synthesis techniques. Furthermore, high-level Esterel Specifications and Concurrent Action-Oriented Specifications are some of the other alternatives which are used to synthesize hardware designs. High-level synthesis from such specifications has also shown potential to raise the abstraction level of HDLs.

Thus, high-level synthesis can be performed using a variety of high-level input specifications. Moreover, different internal representations are used by different synthesis tools to represent these input high-level specifications before converting them into the RTL code. Below, we discuss in brief some of the well-known high-level synthesis techniques differing in terms of their input specifications and/or internal representation used during the synthesis process.

1.2.1 CDFG-Based High-Level Synthesis

Control Data Flow Graphs (CDFGs) are the most common intermediate representation for high-level synthesis. They consist of operation and control nodes with edges for both data-flow dependencies and control flow in a design. The three most important phases of CDFG-based high-level synthesis process are *operation scheduling, resource allocation*, and *resource binding* for which CDFGs provide a suitable representation. During operation scheduling, each operation of a CDFG is mapped to an appropriate control step in which it will be executed. Resource allocation determines the number of various resources (adders, multipliers, registers) that should be used to implement the design. And in resource binding, each operation is bound to a particular resource which will be used to implement it. Thus, during the high-level synthesis process, scheduling, allocation, and binding steps are solved as optimization problems on the CDFG-based data-flow models. Various constraints on the clock period, throughput, or resources of the design make these optimization problems non-trivial.

The scheduling, allocation, and binding phases can be performed in different orders. Performing scheduling before allocation and binding allows the synthesis process to do efficient sharing of various hardware resources based on the compatibility and scheduling of different operations. On the other hand, performing allocation and binding before scheduling allows the synthesis process to schedule various operations based on the interconnection delays. Also, instead of performing each phase separately, two or more phases can also be performed concurrently at the cost of increase in the complexity of the synthesis process. The output of the CDFG-based synthesis process is the RTL code which can be further synthesized to the gate-level netlist using logic synthesis tools [28].

1.2.1.1 HTG-Based High-Level Synthesis

Hierarchical Task Graphs (HTGs), which are hierarchical variants of CDFGs, can also be used as an intermediate representation during a high-level synthesis process. An HTG is a directed acyclic graph having three types of nodes: (1) single nodes which are non-hierarchical nodes and are used to encapsulate basic blocks; (2) compound nodes which have sub-nodes and are used to represent constructs such as if–then–else blocks, switch-case blocks, or a series of HTGs; and (3) loop nodes which are used to represent various loops such as for, do–while. Loop nodes consist of a loop head and a loop tail that are single nodes and a loop body that is a compound node.

SPARK [57], which is a C-to-VHDL high-level synthesis framework, takes the behavioral ANSI-C code (with the restrictions of no pointers and no function recursion) as input and constructs CDFGs as well as HTGs to capture the control flow in the design. The use of HTGs allows global decisions to be taken about various code transformations, thus enabling additional source-level optimizations during the synthesis process. This is because, unlike CDFGs, HTGs maintain global information about hierarchical structuring of the design such as if–then–else blocks and for/while loops. Thus, CDFGs can be used for traversing the design during scheduling whereas HTGs can be used for moving operations hierarchically across large pieces of code without visiting each intermediate node. *SPARK* utilizes both CDFGs and HTGs as intermediate representations during the high-level synthesis of hardware designs.

1.2.2 Esterel-Based High-Level Synthesis

Esterel [15] is a synchronous concurrent language designed to specify control-dominated systems. High-level Esterel Specifications can also be used to generate hardware designs [14, 48]. During the synthesis from Esterel Specifications, designs are generated by first constructing control flow graphs (CFGs) from the input specifications. Each CFG is then converted into a corresponding control dependence graph (CDG), which is a concurrent representation of the sequential program represented by the CFG. CDGs are inherently concurrent and compact (wider and shorter) as compared to sequential, tall, and thin CFGs and enable certain synthesis-time optimizations such as reordering of statements. Esterel-based high-level synthesis utilizes this property of CDGs to generate efficient hardware designs from Esterel Specifications.

1.2.3 CAOS-Based High-Level Synthesis

Another approach of high-level synthesis which has recently been proposed allows a designer to describe a hardware design in terms of guarded atomic actions at a level of abstraction higher that RTL. Such specifications then undergo synthesis to generate the RTL code with automatic handling of the scheduling and synchronization issues, which need to be manually handled in handwritten RTL. We call this specification model CAOS or Concurrent Action-Oriented Specifications. In the CAOS formalism, each atomic action consists of an associated condition (called the *guard* of the action) and a body of operations. In hardware, an action executes only when its associated guard evaluates to *True*, and the body of an action operates on the state of the system [10, 62].

Example

CAOS-based description of GCD (Greatest Common Divisor) design can be written in terms of actions *Swap* and *Diff* as shown in Fig. 1.2.

$$\text{Action Swap}: g_1 \equiv ((x > y) \&\& (y \neq 0))$$
$$x <= y;$$
$$y <= x;$$

$$\text{Action Diff}: g_2 \equiv ((x \leq y) \&\& (y \neq 0))$$
$$y <= y - x;$$

Fig. 1.2 CAOS description of GCD design

The execution semantics of the design shown in Fig. 1.2 is as follows: g_1 and g_2 are the guards of actions Swap and Diff, respectively (x and y are the registers). The swap of the values in the body of action *Swap* occurs only when guard g_1

evaluates to *True*. The subtraction operation $y <= y - x$ in the body of action Diff occurs when g_2 evaluates to *True*. Here, $<=$ denotes the non-blocking assignment operation such that the input values for all such assignments are read before the corresponding operations take place. All non-blocking assignments in the code occur once in a clock cycle and they can all happen at the same time.

Synthesis from CAOS (explained in detail in Chapter 3) generates a scheduler circuit which in each clock cycle dynamically evaluates which actions of a design can be executed concurrently. Thus, a group of operations belonging to an action are considered together for execution in each clock cycle. In contrast, scheduling phase of the typical CDFG-based synthesis generates an efficient static schedule of operations over a sequence of clock cycles. Each operation is individually scheduled during the CDFG-based synthesis from high-level behavioral models.

CAOS model allows the designer to describe the functionality of a design using guarded atomic actions. Each action consists of operations which execute concurrently similar to the real hardware. Thus, CAOS-based descriptions aid in inherent modeling of the hardware designs at a high level and provide a path to automatic synthesis of designs from such models. CAOS-based synthesis has recently been introduced in the EDA industry by *Bluespec Inc. Bluespec Inc.*'s high-level synthesis tool called *Bluespec Compiler* (BSC) takes high-level *Bluespec System Verilog* (BSV) code as input and generates RTL code comparable to handwritten RTL in terms of latency and area. BSV is a prime example of CAOS-based specification language and formalism [10].

1.3 Low-Power Hardware Designs

The average power dissipation in a hardware design consists of the following main components: (i) *Dynamic power* which is caused by the *switching activity* occurring in the design, (ii) *Leakage power* which is the static component of the power dissipation, and (iii) *Short-circuit power* which is caused by long signal transition times. In addition, *Peak Power* of a design is defined as the maximum instantaneous power consumed during the execution of the design [94].

Traditionally, the major concerns of most hardware designers are area, performance, cost, and reliability. Thus, for a high-level synthesis methodology to be successful, it should be able to efficiently address these concerns. In addition, power consumption of the generated designs has begun to be another important criteria affecting the viability of a synthesis process. The two main contributing factors making low power consumption of a design an important factor are the following:

1. The pervasive use of personal computing devices (such as portable systems and multimedia products) and wireless communications systems (such as personal digital assistant) having limited battery life makes power management essential for their prolonged operation.
2. High power consumption (mainly due to rapid increase in chip complexity and clock frequencies) results in excessive heating which is undesirable from cooling, reliability, and packaging considerations and places a limit on the number of transistors that can be integrated on a single chip.

The above-mentioned considerations require that synthesis processes at different levels of abstraction should be power aware. In a top-down design flow, the decisions made at the higher levels of abstraction have more beneficial impact than those made at the lower levels of abstraction. Figure 1.3 shows the power savings that can be achieved using optimization techniques at various levels of abstraction. Optimizations performed at the RTL, gate level, or physical level can achieve around 20–50% power savings for a hardware design. In contrast, high-level power optimizations can lead to 2–20× reduction in the power consumption [95]. This is because at high level better decisions about the structure of a design can be taken which allows the designer to choose a power-optimized architecture for the design. The higher abstraction level allows a simple and intuitive analysis of certain exploits that exist in the model. Discovering these opportunities at the RTL and lower levels requires a much more involved circuit or gate structure analysis. Thus, it has become necessary to develop optimization techniques that target power minimization during a high-level synthesis process.

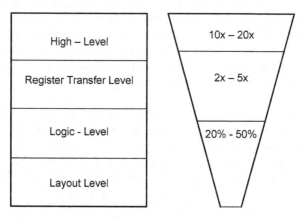

Fig. 1.3 Power savings at various abstraction levels

1.3.1 Power-Aware High-Level Synthesis

During a CDFG-based synthesis process, different phases can target the optimization of area, performance, and power of the generated hardware designs. Scheduling affects the allocation and binding phases, and hence efficient scheduling of the operations of a design is important for producing optimized designs. The problem of scheduling during traditional high-level synthesis using CDFGs can manifest in different forms as per the design requirements. A resource-constrained scheduling problem involves the minimization of the total number of control steps required to execute the operations of a design given the fixed number of each resource. On the other hand, time-constrained scheduling problems involve minimizing the number of required instances of each resource given the fixed number of control steps.

Similarly, for power minimization during CDFG-based synthesis, scheduling under the performance constraints is done. During synthesis, transformations such as pipelining and loop unrolling are used in order to exploit the concurrency in a hardware design while increasing its throughput. This allows the operation of the design at lower clock frequencies and voltages. The binding and allocation phases of the synthesis process also have an effect on the switching activity of the design, thereby affecting its power dissipation. During the binding and allocation phases, power reduction is achieved by performing power-aware allocation and binding of various registers, functional units, and interconnects of the designs. A lot of research work has been done in the area of producing low-power hardware designs from CDFG-based synthesis.

In the CAOS-based synthesis, various area optimizations like resource sharing, common subexpression elimination are applied during the synthesis process. Moreover, scheduling of the actions of a design is done such that in the generated hardware maximal set of actions are executed in each clock cycle. Thus, the performance of a design is also effectively targeted in CAOS-based synthesis. This means that CAOS-based synthesis inherently targets the optimization of the area and performance of a design. However, since such a synthesis process is relatively new as compared to the CDFG-based synthesis, the problem of generating power-efficient hardware from CAOS has not been addressed yet. As mentioned earlier, a viable high-level synthesis process should efficiently address the power minimization problem in addition to meeting the area and performance requirements. Thus, there is a need to augment the CAOS-based synthesis process with power optimization algorithms and techniques.

During CAOS-based synthesis, scheduling of the actions of a design can be used to reduce the number of operations executing concurrently in a clock cycle (thus reducing the peak power) as well as to minimize the switching activity of various signals of the design (thus targeting dynamic power reduction). Also, operations corresponding to the actions which are disabled (whose guards evaluate to *False*) in a clock cycle can be avoided for low power consumption. Thus, various low-power optimizations can be performed during the synthesis of hardware designs from CAOS in order to generate power-efficient RTL code.

In the first part of this book, we consider the CAOS model and perform a complexity analysis of the low-power optimization problems germane to the CAOS-based synthesis process. We also propose various heuristic algorithms which can be successfully used to synthesize power-efficient designs from CAOS. Using the proposed techniques, effects of various low-power techniques on different architectures of a design can be estimated much earlier in the design cycle (at RTL), and appropriate architecture and low-power technique can be selected leading to increase in the overall productivity. This enhances the achieved power savings and aids in easier and faster architectural exploration by avoiding the need to go through the whole power estimation flow up to the gate level for each architectural choice. Furthermore, since such low-power techniques [89] are very commonly used in most real hardware designs, this implies that the implementation of such techniques during CAOS-based synthesis generates designs which represent real designs more closely, as compared to the designs which are generated without using the proposed low-power techniques.

1.4 Verification of Power-Optimized Hardware Designs

Using various power optimization techniques during the synthesis process may result in changing the behaviors of the generated hardware designs. Thus, verification of such designs is needed in order to ensure that they satisfy all the required properties even after they have been optimized for low power consumption. However, ensuring the correctness of a hardware design is the most critical, complex, and time-consuming part of the development cycle. A variety of techniques ranging from simulation to formal verification can be used to approach this problem. Although conventional simulation techniques are effective in the early stages of debugging when the design still contains multiple bugs, such techniques become inadequate as the design becomes cleaner. Moreover, as the complexity of the designs increases drastically the inability of such techniques to scale up makes them ineffective to reason about the correctness of complex designs [37].

Formal verification is widely agreed upon to be a very useful method for proving the correctness of a design. It is a process of using formal mathematical techniques for proving or disproving the correctness of a hardware design with respect to a certain formal specification or property. One of the commonly used approaches for the formal verification of hardware designs is model checking, which is a technique for verifying concurrent systems such as sequential circuit designs and communication protocols. It entails systematic exploration of the relevant states of a design model in order to verify the specified correctness properties. Model checking avoids the construction of complicated proofs and provides counterexample trace when a required property of the system is not satisfied. Apart from these, it also offers several other advantages over the traditional approaches for verification [37].

Model checking is widely employed in the RTL and gate-level models of hardware designs for verification of their desired properties. However, the primary challenge in model checking is that for complex designs it is known to be associated with the state space explosion problem. This problem occurs in designs with many components that interact with each other and make transitions in parallel. This leads to an enormous increase in the number of global states of the design. For this reason, when using the low-level models (RTL and gate-level) of complex designs, application of the model checking-based verification method involves the use of various domain-specific abstraction techniques. This is done in order to reduce the state space required to be explored. The abstraction techniques help develop low-complexity models of the complex designs such that the irrelevant details in the descriptions of the design are ignored while maintaining its desired behavior [37].

1.4.1 Verification Using CAOS

Applying verification techniques to high-level (above RTL) models of a design may avoid the need for using some abstraction techniques since such models already ignore the low-level details which are irrelevant for verifying a design's high-level behavioral properties. Thus, high-level models of hardware designs in the form of

CAOS can be utilized for verification purposes early in the design cycle. CAOS-based models can be used for checking essential properties of the designs as well as for verifying any changes caused in their behaviors due to the use of any power optimization techniques.

Verification of CAOS has not been investigated much in industry or academia. As mentioned in Section 1.2.3, an action in CAOS is composed of a collection of operations. Each action is atomic in the sense that all its operations are executed without being interleaved or interrupted by operations of other actions of the design. We believe that with regard to verification of a design, this clearly aids in reducing its state space since the transitions within a design can be modeled at the higher abstraction level of various actions, instead of individual operations. Thus, CAOS models can be effectively used to quickly verify various correctness properties of hardware designs at a high level.

In the second part of this book, we investigate various verification problems relevant to CAOS-based designs and propose techniques to solve those problems at a level of abstraction above RTL. Those techniques can be used for the verification of various power-minimized designs generated using the techniques proposed in this book. This involves verification of the desired properties of a CAOS-based design as well as comparing behaviors of various implementations of a CAOS-based design differing in their scheduling of actions.

1.5 Problems Addressed

The main focus of this book is to solve the problems of low-power synthesis related to CAOS and verification of the synthesized low-power implementations. In our view, these are the two main problems which have not been targeted yet in the CAOS model and are critical for the success of hardware design methodology based on high-level synthesis from CAOS. In summary, the problems addressed in this book are as follows:

1. Low-power hardware synthesis from CAOS

   a. *What low-power optimizations are appropriate for incorporating in CAOS-based synthesis?* Power-efficient hardware designs can be generated from CAOS by implementing low-power optimizations during the synthesis process. In order to implement suitable low-power optimizations, complexity analysis of low-power problems associated with CAOS-based synthesis is mandatory. In this book, we performed a complexity analysis of dynamic power and peak power problems germane to the CAOS-based synthesis process. Such an analysis allowed the selection of appropriate low-power techniques as discussed below.

   b. *How to reduce dynamic power consumption in designs generated from CAOS?* Dynamic power is a major component of the total power consumption of a hardware design. It can be reduced by minimizing switching activities occurring in various parts of the design. This book proposes various dynamic power reduction techniques which exploit the CAOS model of computation in order to avoid unnecessary computation in a hardware design generated from CAOS. We implemented some of those techniques in *Bluespec Compiler* (which is a high-level synthesis tool based on CAOS), tested them on various realistic designs, and present the corresponding experimental results in this book.

c. *How to reduce peak power consumption in designs generated from CAOS?* Peak power is important from cooling, reliability, and packaging considerations. Minimization of the peak power of a design can be achieved by reducing the number of operations occurring instantaneously in a clock cycle. We propose various peak power reduction techniques which target the restriction of the maximum number of concurrent operations in hardware designs generated from CAOS. We implemented one such peak power reduction technique in *Bluespec Compiler*, tested it on various realistic designs and present the corresponding experimental results in this book.

2. Verification of power-minimized CAOS-based designs

a. *How to enable formal verification of CAOS-based designs?* Formal verification of a CAOS-based design is mandatory in order to ensure that its specification and various implementations adhere to designer's intentions of the design behavior and meet all the correctness requirements even after using any power optimization technique. In this book, we investigate various verification problems relevant to CAOS-based designs and propose techniques to solve those problems at a level of abstraction above RTL. The proposed techniques can be used for the verification of power-minimized CAOS-based designs.

1.6 Organization

As shown in Fig. 1.4, this book is organized as follows:

1. Chapter 2 discusses related research work done in the domain of low-power high-level synthesis and verification of high-level specifications.
2. Chapter 3 provides a detailed explanation of various background topics (such as CAOS-Based Synthesis Process, Power Components of Hardware Designs, Formal Verification) relevant to this book.
3. Chapter 4 presents a complexity analysis of various power optimization techniques related to CAOS-based synthesis process. Both dynamic power problem and peak power problem are analyzed.
4. Chapter 5 proposes various heuristics for peak power and dynamic power reductions that can be applied during the CAOS-based synthesis of hardware designs.
5. Chapter 6 presents a detailed complexity analysis of the problem of scheduling of actions for synthesis from CAOS. The main focus of this chapter is on the

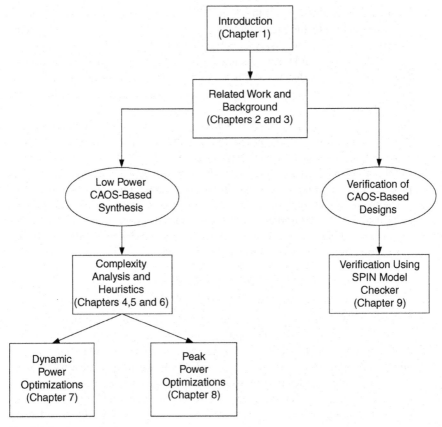

Fig. 1.4 Book organization

optimization of conflicting goals of peak power and latency for CAOS-based hardware designs.

6. Chapter 7 presents how power saving techniques, namely operand isolation and register-clock gating, can be used to save dynamic power in designs generated from CAOS. We propose two algorithms and present the experimental results obtained by applying those algorithms to reduce the power consumptions of some realistic hardware designs.

7. Chapter 8 presents a peak power reduction algorithm which can be applied for peak power reduction of designs generated during CAOS-based synthesis process. Experimental results using some realistic designs are also presented.

8. Chapter 9 presents techniques which can be used for formally verifying essential properties as well as for proving stronger language-containment results for hardware designs synthesized from CAOS models.

9. Chapter 10 summarizes our main contribution and concludes by a discussion on future work.

Chapter 2
Related Work

2.1 High-Level Synthesis

High-level synthesis of hardware designs has been shown to be possible from a variety of high-level HDLs which are used to specify the behavior of the designs at a level of abstraction above RTL. Hardware descriptions written using such HDLs are passed as inputs to high-level synthesis tools to generate the RTL code. Below, we briefly discuss some of the high-level specification languages and corresponding synthesis tools used in the industry and academia.

2.1.1 C-Based Languages and Tools

Since the late 1980s, various techniques for hardware synthesis from many C-like languages have been proposed [45, 49]. Most techniques either use a subset of C or propose extensions to C in order to synthesize the high-level code to the RTL. *Cones* [113] synthesis system from AT&T Bell Laboratories allows the designer to describe the circuit behavior during each clock cycle of a sequential logic. It takes a restricted subset of C (with no unbounded loops or pointers) as input for hardware synthesis. Similarly, *HardwareC* [74] is a C-like language having a cycle-based semantics with support for hardware structure and hierarchy. It is used as an input to *Olympus* synthesis system [46]. *Celoxica's Handel-C* [24] is a superset of ANSI-C with support for parallel statements and rendezvous communication.

SpecC [53, 88], which is a system-level design language based on ANSI-C, has an explicit support for hierarchy, concurrency, state transitions, timing, and exception handling. *Spec*'s synthesizable subset is used to generate hardware designs. *SystemC* [56] is an open-source modeling and simulation language, a subset of which is synthesizable to hardware. It is basically a library of C++ classes that allows RTL as well as high-level modeling using hardware-oriented constructs.

Various high-level synthesis tools which take such C-like HDLs as input have also been proposed [104]. *Forte's Cynthesizer* [52] and *Celoxica's Agility Compiler* [23] are examples of tools which provide synthesis path from high-level *SystemC* models to RTL. *Mentor's Catapult* [82] is a high-level synthesis tool that uses ANSI-C++ to generate RTL code. *PICO Express* by *Synfora* [114] synthesizes

G. Singh, S.K. Shukla, *Low Power Hardware Synthesis from Concurrent Action-Oriented Specifications*, DOI 10.1007/978-1-4419-6481-6_2,
© Springer Science+Business Media, LLC 2010

algorithmic C descriptions of a design into RTL. *NEC's Cyber* [117] tool accepts behavioral descriptions in C or behavioral VHDL and generates corresponding RTL. *SPARK* [57], which is a high-level synthesis framework, takes the restricted ANSI-C code (with no pointers or function recursion) as high-level behavioral input and generates RTL VHDL.

2.1.2 Other Languages and Tools

Apart from C-based tools and languages, synthesis from other high-level specifications has also been proposed. *Esterel* [15], which is a synchronous language, allows the description of a design at high level which can then be synthesized using *Esterel Studio* [50]. *Esterel* supports implicit state machines and provides constructs for concurrent composition, preemption, and exceptions [14, 48]. *Bluespec System Verilog (BSV)*, which is based on Concurrent Action-Oriented Specifications (CAOS), is another such example. It allows the description of a hardware design at a high level in terms of guarded atomic actions, which can then be synthesized to RTL using *Bluespec Compiler* [18].

Optimization and verification techniques for synthesis from C-like languages have been given significant attention in order to produce efficient hardware designs. However, synthesis from CAOS has recently been introduced in the industry and low-power optimizations and verification problems related to CAOS-based high-level synthesis have not been thoroughly investigated. As mentioned earlier, in this book, we investigate and propose various low-power optimization and verification techniques relevant to synthesis from CAOS-based high-level languages such as *BSV*.

2.2 Low-Power High-Level Synthesis

During a high-level synthesis process, information about the data flow and control flow between various operations of a design is used in different phases. For this reason, high-level synthesis from C-like HDLs commonly use Control Data Flow Graphs (CDFGs) as intermediate representations during the synthesis process, and consequently, most research in the area of low-power high-level synthesis is targeted toward the CDFG-based synthesis. In the past, various power optimization techniques targeting the reduction of the dynamic power as well as the peak power of hardware designs during the CDFG-based synthesis have been proposed, some of which are briefly discussed below.

2.2.1 Dynamic Power Reduction

2.2.1.1 Scheduling, Allocation, and Binding

Scheduling of various operations of a design can be exploited for generating power-efficient designs. The problem of resource-constrained scheduling for low power has been addressed in [75, 102]. These approaches use CDFGs to first determine the mobility of various operations based on the ASAP and ALAP schedules. Using the computed mobilities and other relevant factors, priorities are assigned to various operations. Based on the assigned priorities, various operations of the design are then scheduled in each clock cycle such that the power consumption of the design is reduced.

During the allocation phase of a high-level synthesis process, functional units and registers are allocated to the design, whereas in the binding phase, operations and variables of a design are mapped to the allocated functional units and registers, respectively. References [30, 78, 87, 93] present techniques targeting low-power reduction during allocation and binding phases. Raghunathan and Jha [93] present an allocation method for low-power high-level synthesis, which selects a sequence of operations for various functional units such that the overall transition activity of the design is reduced.

Murugavel and Ranganathan [87] present an algorithm targeting the minimization of the average power of a circuit during the binding phase of high-level synthesis process using game-theoretic techniques. In that work, binding of the operations to the functional units is done such that the total power consumption is minimized. Lakshminarayana et al. [78] present an efficient register-sharing and dynamic register-binding strategy targeting power reduction in data-intensive hardware designs. The experiments demonstrate that for a small overhead in the number of registers, it is possible to significantly reduce or eliminate spurious computations in a design. The proposed strategy handles this problem by performing register duplication or an inter-iteration variable assignment swap during the high-level synthesis process.

Chen et al. [30] target low-power design for FPGA circuits. It presents a simulated annealing engine that simultaneously performs scheduling, resource selection, functional unit binding, register binding, and datapath generation in order to reduce power. Chen et al. [30] also propose a MUX optimization algorithm based on weighted bipartite matching to further reduce the power consumption of the design.

2.2.1.2 Power Management

Power management refers to techniques that involve shutting down parts of a hardware design that are not being used for power savings. This can be done by disabling the loading of a subset of registers based on some logic [29, 43, 69, 86]. In [69], authors introduce a power management technique called *Pre-computation* to improve such power management during the high-level synthesis process. Monteiro et al. [69] present a scheduling algorithm which maximizes the shut-down period of execution units in a design. Given a throughput constraint and the number of execution units available, the algorithm schedules operations that generate controlling signals and activates only those modules whose result is eventually used.

Another well-known *power management* technique is *Operand Isolation* (also known as *signal gating*) which avoids unnecessary computations in a design by gating its signals in order to block the propagation of switching activity through the circuit. Chattopadhyay et al. [29] discuss automation of operand isolation during

ADL (Architecture Description Languages)-based RTL generation of embedded processors. In [86], a model is described to estimate power savings that can be obtained by isolation of selected modules at RTL.

Dal et al. [43] define power island as a cluster of logic whose power can be controlled independent from the rest of the circuit, and hence can be completely powered down when all of the logic contained within it is idling. Dal et al. [43] propose a technique that eliminates spurious switching activity and the leakage power dissipation in a circuit through the use of power islands. The technique first schedules the design in order to identify the minimal number of resources needed under the given latency constraints. After scheduling, the functional unit binding is done based on the life cycle analysis of all the operations. The scheduling and binding steps are followed by a partitioning phase which uses a placement algorithm that performs partitioning such that the components with maximally overlapping lifetimes are clustered together and assigned to the same power island. After the partitioning phase, register binding is performed such that both the total and the average active cycles of registers are minimized.

2.2.1.3 Power-Efficient Floor Planning

Scheduling and binding phases of a high-level synthesis process can significantly affect the switching activity and topology of the interconnects (wires, buffers, clock distribution networks, multiplexors, and busses) in the final design. References [91, 120] are based on the idea that it is important to consider the impact of physical design during high-level synthesis. Zhong and Jha [120] propose a technique which first evaluates the power consumption in the steering logic and clock distribution network in addition to data transfer wires using floor-planning information. Then, an ASAP schedule of the design is determined followed by an interconnect-aware binding in order to optimize area and power of the design. The proposed binding technique tries to localize data transfers by ensuring that as many RTL neighbors as possible are also the physical neighbors, especially those that exchange data with each other. After each binding move, the behavior is re-scheduled if required. The technique proposed in [91] considers the effect of scheduling and binding on the floor planning. The authors propose two parallel algorithms for simultaneous scheduling, binding and floor planning, both of which are based on the simulated annealing algorithm.

2.2.1.4 Other Work

Some other low-power high-level synthesis work includes [27, 70, 115]. Reference [70] presents a high-level synthesis system for targeting reduction of power consumption in control flow-intensive and data-dominated circuits. The system uses an iterative improvement algorithm to take into account the interaction among the different synthesis tasks for power savings. Chandrakasan et al. [27] present a high-level synthesis system for minimizing power consumption in application-specific datapath-intensive CMOS circuits using a variety of architectural and computational transformations. The proposed approach finds computational structures that result in the lowest power consumption for a specified throughput given a high-level algorithmic specification. Uchida et al. [115] propose a thread partitioning algorithm for low-power high-level synthesis systems. The algorithm divides parallel behaving circuit blocks (threads) of a design into subparts (sub-threads) such that gated clocks can be applied to each part for power savings.

2.2.2 Peak Power Reduction

As proposed in [84, 85, 96, 100, 101], high-level synthesis of hardware designs can also be used to target the peak power reduction of the generated designs. Raghunathan et al. [96] propose a technique which enhances the scheduling phase of a high-level synthesis process to handle the constraints on the transient power characteristics of a design. The techniques insert data monitors at appropriate parts of a hardware design. During the execution of the design, such monitors are used to select (based on the input values) if certain operations can be allowed to execute together in a given clock cycle under the specified transient power constraints.

In [101], an ILP (Integer Linear Programming)-based scheme for high-level synthesis of low-power applications is presented. The authors present an ILP-based model for latency-constrained scheduling that minimizes the number of resources, peak power consumption, and peak area for the design. An extension of [101] is presented in [100]. It proposes an ILP model as well as a modified force-directed scheduling (MFDS) heuristic that minimizes the peak power of the design while satisfying its timing constraints. The results obtained by the heuristic are approximate as compared to the results obtained by the ILP methods, which are optimal. However, the proposed heuristic-based algorithm is faster than the ILP methods.

Mohanty and Ranganathan [84] propose a framework for simultaneous reduction of total energy, average power, peak power, and peak power differential during high-level behavioral synthesis of deep sub-micron and nanometer designs. A new parameter called "Cycle Power Profile Function" (CPF) is defined which captures the transient power characteristics of a design as a weighted sum of mean cycle power and mean cycle differential power. Mohanty and Ranganathan [84] present a datapath scheduling algorithm which attempts to minimize the CPF. The CPF scheduler assumes several types of resources at each voltage level and number of allowable frequencies as resource constraints. Energy savings are achieved by using the energy-hungry resources operating at reduced voltages to the maximum extent. The loss in performance is compensated by maximizing utilization of lesser energy-consuming resources operating at higher voltages so that they can be operated at higher frequencies. Mohanty et al. [85] also present a similar work where the authors propose an ILP formulation for the minimization of CPF during datapath scheduling.

2.2.3 Summary – Low-Power High-Level Synthesis Work

Dynamic power is usually the most important component of the total power consumption of a design, and thus its reduction is targeted during most power-aware high-level synthesis processes. As we move toward the deep sub-micron levels, leakage power minimization is also becoming a critical issue. Furthermore, peak power minimization is also as important a goal as dynamic/leakage power minimization. Thus, simultaneous reduction of dynamic, leakage, and peak power components during a high-level synthesis process is favorable.

During high-level synthesis, most power savings are achieved by optimizations performed during the scheduling phase. Power-efficient allocation and binding further aids in increasing the achieved power savings. Physical level information can also be utilized during these phases for power reduction. Thus, during a high-level synthesis process, a wide variety of techniques can be applied for generating power-efficient hardware designs. Moreover, various analysis techniques such as ILP formulation and Game Theory can be applied during the synthesis process in order to arrive at an efficient solution for the low-power minimization problem.

The above-mentioned power minimization techniques are relevant to CDFG-based synthesis. In future, high-level synthesis using CAOS is expected to gain more acceptance in the EDA industry. Thus, more techniques targeting power minimization during this approach of synthesis are required. This book presents strategies targeting the reduction of peak power and dynamic power in the hardware designs generated from CAOS.

2.3 Power Estimation Using High-Level Models

In order to take efficient power reduction decisions during a high-level synthesis process, it is important to find ways and techniques for estimating the power consumption of a hardware design at a level of abstraction above RTL. Though high-level power estimation using CAOS is not the topic of research presented in this book, for the sake of completeness of the discussion, below we present some related work done in the area of power estimation using high-level model of hardware designs.

Ahuja et al. [8] present a methodology, which utilizes the power estimation knowledge to guide power reduction. The premise to their work is to first provide an accurate and efficient power estimation framework at higher abstraction level and then utilize this information to guide the power reduction algorithm while generating the hardware RTL. On the power estimation side, they present mainly three different approaches: (a) reusing the RTL power estimation frameworks at higher level, (b) providing characterization-based power models to facilitate the power estimation at a high level, (c) utilizing various verification collaterals such as assertions and properties to speed up the design flow for power estimation.

The idea of reusing RTL power estimation frameworks in high-level synthesis-based design methodologies comes from the fact that the biggest bottleneck for power estimation at RTL is the processing of large simulation dumps capturing various signal activities. In [6], Ahuja et al. provide a rationale for their approach and show that based on modeling style at high level, the speedup in power estimation time can be achieved. On a variety of benchmarks, they experimentally showed that the range of speedup for high-level power estimation process reaches up to 15 times as compared to RTL power estimation techniques. The error or loss of accuracy using their proposed methodology was less than 10% with respect to lower level power estimates. The benchmarks used in their methodology include a processor model (VeSPA) and DSP algorithms (such as FFT, FIR).

Ahuja et al. [2, 5] present characterization-based power estimation methodology utilizing GEZEL [55]-based co-simulation environment. In this environment, a hardware modeled as FSMD (Finite State Machine with Datapath) in GEZEL can be co-simulated with existing processor models such as ARM. The power model proposed in this case study is a regression-based power model which is learnt through various statistical test data. The purpose of using a regression model is to utilize various states of FSMD as regression variables as opposed to input- and output-based characterization in previous approaches. This gives a better visibility on power estimation because this approach helps in relating activity of the states of FSMD to power consumption. Authors present the theory and rationale behind the model and their approach. This power model is verified on a variety of benchmarks modeled in GEZEL and co-simulated with ARM processor's simulation model. Furthermore, the work is aided by working on different technology nodes (130 and 90 nm) and an elaborated discussion is provided on how to improve the estimation process in the presence of lower level power reduction techniques such as clock-gating and power gating. Error in the analysis was less than 5% in the presented experiments.

The estimates from the power estimation methodology/tools are not accurate if they are not supplied with good representative test vectors. Ahuja et al., in another work in [3, 4], present a methodology to utilize verification collaterals to enhance the accuracy of power estimation at high level. They use Esterel [50]-based high-level modeling framework to model control-intensive designs such as a power state machine controller. They present how, for a controller from high level, various properties using assertions can be exercised using the formal verification framework inbuilt in the Esterel Studio. They also show how to write negative properties to help in creating counter cases to provide reachability of a state, transitions from a state, and staying in a particular state of the controller using invariants. These counter cases can be synthesized to test vectors, and data vectors are used as random test vectors while control test cases are created by utilizing properties. These comprehensive sets of test vectors provide the behavior of the controller of the design for different states of the design at high level. They measure the power for each state and also the associated valid transitions and finally utilize these numbers during the high-level simulation of the model. The overall idea is to utilize the best of both the worlds, i.e., simulate at high level, while using power estimation-related knowledge by actually exercising the RTL model to different test scenarios.

Ahuja et al. [9] present a power reduction technique based on clock-gating and sequential clock-gating from high-level model descriptions. The authors use a tool C2R [1, 22] for their case study and show how from the ANSI-C description of a design, a clock-gated RTL can be auto-generated from the power reduction perspective. They show how various granularities of clock-gating can be exercised from the high-level description itself. This helps in finding out the opportunities which lower level tools cannot achieve. In the high-level synthesis context, this is especially important because a lot of block-level clock-gating decisions are done at the high level as the design complexities are increasing. Thus, it is important to control this kind of coarse grain power reduction directly from the ANSI-C description. The work in [9] presents how various pragmas related to clock-gating can be introduced in the design which can then be recognized by the high-level synthesis tool. They also show how to correctly capture the designer's intent on power management using clock-gating with these pragmas. In their analysis, they also show how various lower level implementation features can be facilitated in high-level synthesis flow. This includes a way to describe integrated clock-gating cell (ICGC) in the generated RTL. ICGC is generally provided by the library vendors. The presented algorithm considers finding out common enables for the registers such that minimum number of ICGCs are inserted on the clock-paths across the blocks. The other indirect benefit of such an approach is the insertion of ICGCs closer to the root of the clock, otherwise there would be lot more ICGCs inserted toward the leaf of the clock tree, which in turn could cause additional power wastage in the clock tree.

In order to extend the clock-gating-based power reduction, Ahuja et al. [7] also show how sequential clock-gating opportunities can be identified at higher abstraction levels. Generally the difficulty to implement the sequential clock-gating comes from two facts: (1) if the designers are trying to find such opportunities they might change the behavior of the design, (2) if the tools are used to facilitate such an optimization they generally work at the netlist level and are used to automatically find out the observability do not care or stability conditions. This hinders the exploration of such an optimization at high level. Ahuja et al. [7] present a model checking-based technique first showing how to model this power reduction feature using the properties (the authors show that the dependency among registers can be captured in a property) and then supplying it to a model checker. If the model checker passes that property then those registers become the possible candidates for sequential clock-gating. These conditions are put as always true properties, which is a strong condition but this helps in reducing the chances of bugs later in the design stage. This approach addresses the two problems mentioned earlier: (1) ease of verification and (2) since the properties are verified on the behavioral model, there is no need to go to the netlist level. This work is further extended for use by a high-level synthesis tool, showing that once such a technique is adopted at a high level, the power savings can be higher than clock-gating for certain applications.

Clock-gating and operand isolation are two techniques widely used to reduce the power consumption of state-of-the art hardware designs. Both approaches basically follow a two-step procedure: first, they statically analyze a hardware circuit to determine irrelevant computations; second, all parts which are responsible for these computations are replaced by others that consume less power in the average case, either by gating clocks or by isolating operands. Brandt et al. [19] define the theoretical basis for adoption of these approaches in their entirety. They show how irrelevant computations can be eliminated using their approach. They present passiveness conditions for each signal x, which indicates that the value currently carried by x does not contribute to the final result of the system. After showing how their theory can be generally used in the context of clock-gating and operand isolation, a classification of many state-of-the-art approaches is performed, and it is shown that most of the approaches in the literature are conservative approximations of their general setting.

Ahuja et al. [8] present a methodology to automatically find the clock-gating opportunities by performing a system-level simulation of the design. To find such opportunities, authors present a system-level power estimation model which utilizes certain properties inbuilt in the clock-gating logic. They show how technology-dependent and -independent models help in measuring the efficacy if clock-gating would have been applied at the high level itself. This information is utilized to apply the clock-gating at an appropriate place. The power model is integrated with untimed transaction-level test/application environment, which is shown to be 2–3 orders of magnitude faster than the state-of-the-art RTL power estimation techniques. This helps in analyzing more power reduction opportunities and helps in putting the appropriate granularity of clock-gating. This can also serve as a basis of applying estimation-guided reduction for other optimizations such as sequential clock-gating, operand isolation, memory gating. Most of the current approaches rely on using the reduction algorithm based on the architect's or designer's knowledge, while this approach distinguishes itself by finding the possible power savings by utilizing fast and accurate power reduction models.

Lakshminarayana et al. [77] present a methodology to quickly explore various hardware configurations and also to obtain relatively accurate design matrices for each configuration. They use high-level synthesis tool, C2R, to quickly generate RTL descriptions of the hardware. They show how high-level synthesis languages give a greater degree of freedom to do selective optimizations by means of inserting the so-called directives into the behavioral code (also known as restructuring). They present case studies to develop or modify behavioral IP descriptions and use standard FPGA boards to profile the IP in a very short time. The differences in the measured and the actual IP design matrices are not significant as one is more concerned with the relative differences among various configurations. A variety of compute-intense benchmarks like AES are used to demonstrate how platform-specific optimizations as well as high-level microarchitectural optimizations can be done using a commercial high-level synthesis tool along with Xilinx Spartan/Virtex boards and Xilinx EDK design suite [119]. The results presented in this chapter show how various architectures in hardware software codesign flow can be chosen keeping energy efficiency in mind. The authors also show how they reduced the design cycle time to reach the optimal results.

2.4 Verification of High-Level Models

Model checking [35–37, 92] was proposed as a verification technique more than two decades ago. Initially, state transition graphs were used for model checking along with efficient graph traversal techniques. A lot of research in model checking is targeted at reducing the state space of the model to target the associated state explosion problem, for which various abstraction techniques are used. Around 1990, techniques that used symbolic state space exploration were proposed [42, 90]. In symbolic model checking, a breadth-first search of the state space is done using BDDs (Binary Decision Diagrams) [20]. However, even in BDD-based symbolic model checking, the problem of state space explosion manifests itself in the form of unmanageable large BDDs which increases the complexity of their manipulation. This problem can be addressed by Bounded Model Checking (BMC) [38], which uses a SAT solver rather than BDD manipulation techniques and has been used to successfully verify large hardware designs.

High-level models of complex designs can also be used to simplify the associated verification problem (as well as shorten the design time). Such models already abstract out irrelevant behavioral details, reducing the state space of the design to be explored. *SystemC* [56], *BSV* [18], *Esterel* [15], *SpecC* [53, 88], etc., are examples of HDLs which can be used to describe such high-level models of hardware designs.

2.4.1 SpecC

SpecC language is based on ANSI-C and contains various constructs for describing state machines, concurrency, and arbitrary-length bit vectors. Verification of *SpecC* models is explored in [68, 98]. Jain et al. [68] describe how SAT-based predicate abstraction can be used to verify a concurrent *SpecC* system description. The abstractions in this work preserve the bit vector semantics of *SpecC* and support all *SpecC* bit-vector operators. A low-level design may also be added which is also abstracted using SAT-based predicate abstraction. Sakunkonchak et al. [98] propose a technique which helps in verifying the synchronization of events in *SpecC* using encoding of data in the form of difference decision diagrams. However, the technique proposed in [98] does not support hardware-level constructs.

2.4.2 SystemC

Verification of *SystemC* models has also been investigated [47, 73, 81]. In [47], authors describe the use of BDDs for reachability analysis of *SystemC* gate-level models. For this, the proposed technique converts such models into BDDs. Man [81] presents a method which is used to check equivalence between *SystemC* designs by converting those designs into BDD representations. Both these techniques use low-level models of designs for verification. Kroening and Sharygina [73] present a technique that handles both hardware and software parts of a system description written in *SystemC*. For better performance, the proposed technique automatically partitions the system description into synchronous (hardware) and asynchronous (software) parts, which are then verified.

2.4.3 Other Work

A significant amount of research work has been done targeting the verification of hardware and software models written in C-like languages [26, 33, 34, 40, 67, 71]. In [71], authors apply SAT-based predicate abstraction to the equivalence checking problem on ANSI-C programs. However, the technique does not support concurrency. Other techniques using SAT-based decision procedure are proposed in [26, 40, 67]. Clarke and Kroening [33] describe an algorithm to verify a hardware design given in Verilog using ANSI-C program as specification. In that work, SAT-based BMC is used in order to reduce the equivalence problem in a bit vector logic decision problem. Clarke et al. [34] propose an algorithm that checks behavioral consistency between a ANSI-C program and a circuit given in Verilog using BMC. The program and the circuit are translated into a formula that represents behavioral consistency, which is checked using a SAT solver. The approach is restricted to sequential ANSI-C programs. Also, the proposed methods are unable to handle large designs since no abstraction techniques are employed.

Other research work related to verification using high-level models include [16, 25, 31, 32]. In [25], authors propose a technique for verifying concurrent (message-passing) C programs against safety specifications. The authors propose a compositional framework which combines two orthogonal abstraction techniques (predicate abstraction for data and action-guided abstraction for events) within a counterexample-guided abstraction refinement (CEGAR) scheme [72]. Clarke et al. [32] present techniques to improve predicate abstraction for verification of hardware/software systems. Authors propose an algorithm which is based on localization reduction. The proposed algorithm helps to avoid the potential blowup of the size of abstract models when verifying large control-intensive systems. This is achieved by selectively incorporating state machines that control the behavior of the system under verification. In [31], authors present a modeling and high-level verification methodology based on Petri Net model. It targets the verification of data flow graphs for DSP algorithms by transforming such graphs into Petri Net models. Bingham et al. [16] discuss a methodology used on an industrial hardware development project at Intel to validate various cache-coherence protocol components by using a high-level model (HLM) written in a guarded command language, Murphi [44], for model checking purposes. The HLM is then used as a checker during dynamic (i.e., simulation-based) validation of the microprocessor RTL code to find design bugs.

2.4.4 Summary – High-Level Verification Work

In order to handle the increasing complexity of hardware designs, the use of high-level modeling languages has been gaining importance. Verification of such models aids in creating correct specifications early in the design cycle. In the past, various techniques for verification of C-like models of hardware designs have been proposed. CAOS has been recently proposed for efficient high-level modeling of hardware designs. Verification of CAOS-based model is still in its infancy and not much research has been done in this domain.

In the second part of this book, we investigate various formal verification problems relevant to CAOS-based designs and propose techniques to solve those problems (at a level of abstraction above RTL) using the SPIN model checking tool [64]. The proposed techniques can be used for the verification of power-optimized CAOS-based designs early in the design cycle.

Chapter 3
Background

3.1 CDFG-Based High-Level Synthesis

A CDFG-based high-level synthesis takes a behavioral specification of a design as input and generates the corresponding RTL code. Apart from the specification, other inputs to a high-level synthesis process can be an optimization function, design constraints, and a module library representing the available components at RTL. The goal of the synthesis process is to generate an RTL design that implements the specified behavior while satisfying the design constraints and optimizing the given cost function.

The synthesis process comprises of multiple tasks. Depending on the synthesis tools, these tasks can be performed in different orders, or one or more tasks can be combined into one. Moreover, the tasks can also be performed iteratively to converge on the desired solution. In general, the following tasks are performed during a CDFG-based synthesis process:

1. *Lexical Processing* – This task involves parsing of the high-level description of the design into an internal representation which is then further processed.
2. *Optimizations* – During the synthesis process, various optimizations are performed on the high-level description. These include optimizations like parallelism extraction, loop unrolling, common subexpression elimination, constant folding, and other program transformations.
3. *Control/Dataflow Analysis* – This refers to the generation of a CDFG corresponding to the high-level description of the design. For this, the various input/output signals and the operations of the design are analyzed for figuring out their data dependencies.
4. *Scheduling* – Scheduling refers to assignment of various operations of the CDFG into specific clock cycles. To generate a schedule, various data dependencies, latencies of the functional units in the library, and constraints are taken into account.
5. *Resource Allocation* – This task determines the set of functional units required to implement the design. Scheduling and resource-binding phases can also affect this allocation of resources.

G. Singh, S.K. Shukla, *Low Power Hardware Synthesis from Concurrent Action-Oriented Specifications*, DOI 10.1007/978-1-4419-6481-6_3,
© Springer Science+Business Media, LLC 2010

6. *Resource Binding* – This refers to the assignment of various values and operations of the design to specific instances of registers and functional units (from
the library), respectively. For optimized utilization of various functional units,
resource-sharing can also be performed.

3.2 Concurrent Action-Oriented Specifications

In the *CAOS* formalism, the behavior of a hardware design is described using a
collection of guarded atomic actions at a level of abstraction above RTL. Each action
consists of an associated condition (called the *guard* of that particular action) and
a body which operates on the state of the system. An action executes only when its
associated guard evaluates to *True*. An action is atomic in the sense that all its operations occur together without being interleaved by the operations of the other actions
of the design. Moreover, various operations of an action execute concurrently like
the operations in the real hardware and hence such a CAOS-based description of a
design is intuitive.

Using CAOS, the state of the design can be explicitly defined by the designer
in the high-level description of the system or, in other cases, can be inferred from
that description. *Bluespec Compiler* [10] is an example of CAOS-based synthesis
tool. In *Bluespec*, there is no need to infer the state because the designer explicitly
instantiates all the state elements of the system (such as registers, FIFOs, memories).
This model then undergoes synthesis to generate the RTL code [10].

An *action A* in the concurrent action-based description of a design can be written
as follows:

$$\text{Action } A : g(s) \rightarrow \{$$
$$s_1 = b_1(s, i);$$
$$s_2 = b_2(s, i);$$
$$s_3 = b_3(s, i); \}$$

Here, $g(s)$ is the guard associated with action A. The symbol s represents the
set of state elements of the design such that $s_1, s_2, s_3 \in s$. The body of the action
contains three statements of the form $s_j = b_j(s, i)$, where $b_j(s, i)$ computes the
subset of the next state of the system using the current values of the elements of s
and the current input i.

3.2.1 Concurrent Execution of Actions

In a very simple CAOS-based model of a hardware design, among all the actions
that are enabled (guards evaluated to *True*) in a clock cycle, only one of them can
be selected for execution. In such a model, execution proceeds sequentially with
only one action being executed in each clock cycle and execution stops when no

guard evaluates to *True*. However, such a model is undesirable from a latency point of view. To improve the latency of a design, two or more non-conflicting actions can be allowed to execute concurrently in the same clock cycle. But for correctness purposes, atomicity of the actions adds a constraint that such a concurrent execution of multiple actions must correspond to at least one sequential ordering of their execution, in which only one action is executed in each clock cycle. Thus, in order to generate appropriate scheduling and control logic that maintains the atomicity of various actions, synthesis from atomic actions involves constructing (at compile time) a single sequential ordering consisting of all the actions of a design.

In hardware, multiple actions can execute concurrently, thus exploiting maximum parallelism present in the design. The system execution stops when no *guard* evaluates to *True*.

3.2.2 Mutual Exclusion and Conflicts

Actions having mutually exclusive guards cannot be scheduled in the same clock cycle. Also, actions conflicting with each other cannot be scheduled in the same clock cycle. A simple example of a conflict is two actions each having an operation updating the same state element. In case two or more conflicting actions are enabled in the same clock cycle, a notion of priority is used to decide which of those actions should be executed in that cycle. A higher priority action is always chosen for execution over all the other lower priority conflicting actions.

3.2.3 Hardware Synthesis

CAOS model is behaviorally higher in abstraction because it supports automatic handling of concurrency and synchronization issues (such as due to updates on the shared state elements) via atomic actions. Hardware synthesis using CAOS can be achieved by implementing each *guard* and *body* as a combinational logic and synthesizing a control circuitry for appropriate scheduling and data selection.

Figure 3.1 shows the schema for a translation from the actions into the hardware. The circuit shown is generated in the concurrent action-based synthesis flow. The guards (g's) and update functions (b's) are computed for each action using a combinational circuit. The scheduler is designed to select a maximal subset of applicable actions under the constraint that the outcome of a scheduling step can be explained as atomic firing of actions in some order. In each time slot, the *Control Circuit* selects as many actions as possible to execute and updates the current state with the resulting values of those actions. Thus, there exists an almost direct translation from the action-oriented specification of the design to its hardware. Note that the synthesis is done without any constraints on the resources of the design. The CAOS-based description of a design has been shown to generate efficient RTL code comparable to hand-coded RTL Verilog [10, 62, 97].

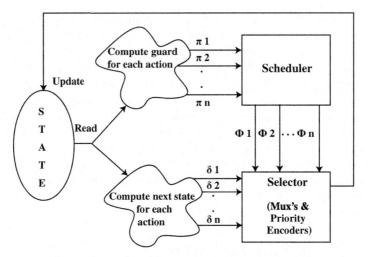

Fig. 3.1 Synthesis from concurrent action-oriented specifications [10]

3.2.4 Example

Figure 3.2 shows an example of a description of a LPM (Longest Prefix Match) module using *Bluespec System Verilog* which is based on CAOS. The example provides an idea of the execution semantics of CAOS [10, 108]. LPM module is used in Internet routers to determine the output port to which a particular packet should be forwarded. It takes 32-bit IP addresses, looks up the destination (32-bit data) for each IP address in a table in a memory, and returns the destinations. One of the

Action 1 (Input):
$(true) \rightarrow \{$
    IPaddr = fifo1.deq();
    token = compBuffer.getToken();
    fifo2.enq(token,IPaddr[15:0],state0);
    RAM.readRequest(baseAddr + IPaddr[31:16]);}

Action 2 (Complete):
$(isLeaf(d = RAM.readResult())) \rightarrow \{$
    (token, IPaddr, s) = fifo2.deq();
    compBuf.done(token, d);
    RAM.readAck(); }

Action 3 (Circulate):
$(!(isLeaf(d = RAM.readResult()))) \rightarrow \{$
    (token, IPaddr, s) = fifo2.deq();
    fifo2.enq(token, IPaddr, s + 1);
    RAM.readRequest(computeAddr(d,s,IPaddr));
    RAM.readAck(); }

Fig. 3.2 Circular pipeline specification of LPM module design using concurrent actions

possible implementations of a LPM module is in the form of a Circular Pipeline as shown in Fig. 3.2.

In the above example, *Action 1* represents the *Input* stage which takes an IP address *IPaddr* from *fifo1* and a token from the *Completion Buffer*. It then places a tuple (token, IPaddr[15:0], state0) into *fifo2* and enqueues a memory request using the 16 bits of the IP address. Based on the memory response *d* and the first tuple (token, IPaddr, s) in fifo2, either the *Completion (Action 2)* or the *Circulate (Action 3)* stage executes. If the lookup is done, the *Completion* stage forwards the tuple containing the memory response *d* and the token to the *Completion Buffer*; otherwise, the *Circulate* stage places the tuple (token, IPaddr, s+1) into *fifo2* and launches another memory request.

The module is pipelined and the results are returned in the same order as requests. The memory is also pipelined and has a fixed latency of L cycles. Since IP addresses need varying numbers of memory references, each IP address goes around the pipe as many times as the number of needed memory references. This means that the requests finish out of sequence. Thus, tokens from the *Completion Buffer* are used to make sure that the results arrive in the right order.

After the parts of an action corresponding to combinational logic have executed, all the other parts of an action will occur in parallel in order to achieve maximum parallelism. For example, in *Action 1 (Input)*, after IP address and token are fetched (combinational parts), the other two parts of the actions which place a tuple into *fifo2* and enqueue a memory request will occur in parallel. Similarly in the *Action 3 (Circulate)*, the portion of the action that places a tuple in *fifo2* enqueues a memory request and acknowledges that the reading of result will be executed in parallel after the part that dequeues the tuple from *fifo2* has executed. This kind of parallelism is quite common in hardware design. However, the use of concurrent actions (multiple actions executing in the same time slot) in the above example reflects the additional amount of parallelism that can be exploited through synthesis from the action-oriented specification style.

The example illustrates that using CAOS relieves designers from worrying about global coordination, thus allowing them to focus on the much simpler task of local correctness. Larger examples of a similar kind may be found in [76, 83].

3.3 Power Components

3.3.1 Average Power

The average power dissipation in a hardware design consists of the following three basic components [94]:

1. *Dynamic Power* – It is related to switching activity occurring in the design. Dynamic power can be expressed as

$$\text{Dynamic power} = kCV^2 fp \tag{3.1}$$

where

> k is a constant.
> C is the overall capacitance which is charged and discharged.
> V is the supply voltage of the design.
> f is the switching frequency of the component.
> p is the switching probabilities of the signals.

Dynamic power is the dominant source of power dissipation in a hardware design and can be reduced during the high-level synthesis process by minimizing the switching activities in the generated hardware.

2. *Leakage Power* – It is the static component of the power dissipation and occurs due to spurious currents in the non-conducting state of the transistors. This component of power can be reduced by using the technique of power gating during synthesis. However, insertion of power-gates might increase latency, and therefore correct grouping of circuit elements and a predictive scheme for triggering power-gating are needed.

3. *Short-Circuit Power* – It occurs due to the stacked P and N devices in a CMOS logic gate that are in the ON state simultaneously and can be minimized by reducing the signal transition times. It is hard to reduce this component of power through synthesis-time optimizations.

3.3.2 Transient Characteristics of Power

Transient characteristics of a circuit refer to its peak power and power differential. Peak power is defined as the maximum instantaneous power (due to switching activity) consumed during the execution of a design. It is important due to packaging and cooling considerations and also affects the I–R drop and electromigration of the power supply network on the chip. Peak power of a design can be reduced by minimizing its peak switching activities. Power differential determines the noise caused due to inductive ground bounce. The use of deep sub-micron technologies makes transient power management in a design as critical as the average power dissipation.

3.3.3 Low-Power High-Level Synthesis

High-level synthesis involves several subtasks, including transformations, module selection, clock period selection, scheduling and resource binding, and RTL circuit generation. All these subtasks significantly effect the power consumption of a design, and their efficient execution can lead to large power savings. Among the high-level synthesis subtasks, behavioral transformations (loop transformations, algebraic transactions, strength reduction, word-length reduction, etc.) enable supply voltage reduction and minimization of switched capacitance. Transformations like pipelining and loop unrolling improve the throughput of a design by exploiting concurrency. This enables the design to be operated at lower clock frequencies, and thereby at lower voltages. The lower supply voltage reduces the power dissipation of the design.

The effect of scheduling on average power consumption is complex and is related to the other high-level synthesis subtasks such as module selection and resource sharing. Scheduling can be used to enable selection of modules which in turn can be used to minimize switched capacitance and also provide opportunities for supply voltage scaling. Scheduling can also be used to enable resource sharing by ensuring that correlated variables can share the same resources and exploiting regularity to minimize interconnect power. Scheduling also affects the peak power consumption of design.

The choice of clock period used during scheduling also affects power consumption. While a large clock period results in resource chaining which results in large glitching power, a very small clock period results in large power consumption in the clock network and registers. Hardware allocation also has an effect on the switching activity and thereby on the power dissipation. Thus, decisions taken during high-level synthesis have a far-reaching impact on the power dissipation of the resulting hardware.

3.4 Complexity Analysis of Algorithms

3.4.1 NP-Completeness

In complexity theory, the class *NP* consists of problems for which there exists a non-deterministic algorithm whose running time is polynomial in the size of the input. *NP-hard* and *NP-complete* are the two classes of problems that contain numerous important problems which are the hardest problems in *NP*. A problem is said to be *NP-hard* if every problem in *NP* is polynomially reducible to it. A problem which belongs to *NP* and is *NP*-hard is said to be *NP-complete* [54]. It is widely believed that *NP*-complete problems do not have polynomial time algorithms. Therefore, heuristic solutions which run in polynomial time and which produce near-optimal solutions are sought for such problems.

A *heuristic* for a combinatorial optimization problem is a polynomial time algorithm that produces a feasible, but not necessarily optimal, solution to all instances of the problem.

3.4.2 Approximation Algorithm

For any $\rho \geq 1$, a *ρ-approximation algorithm* for a combinatorial optimization problem is a heuristic that produces a solution which is within a factor ρ of the optimal solution value. To further clarify this notion, suppose the optimal solution value for an instance I of an optimization problem is denoted by OPT(I). For each instance I of a *minimization* problem, a ρ-approximation algorithm produces a solution whose value is *at most* ρ OPT(I). For each instance I of a *maximization* problem, a

ρ-approximation algorithm produces a solution whose value is *at least* OPT(I)/ρ. A ρ-approximation algorithm is also referred to as an algorithm that provides a *performance guarantee* of ρ.

For any *fixed* $\varepsilon > 0$, a *polynomial time approximation scheme* (PTAS) for a combinatorial optimization problem provides a performance guarantee of $1+\varepsilon$. Thus, by an appropriate choice of ε, a PTAS can produce solutions that are arbitrarily close to the optimal value. As can be expected, the running time of a PTAS increases as the value of ε is decreased.

For some combinatorial problems, it is possible to design algorithms whose running times are polynomial in the size of the problem instance and the *maximum* value that occurs in the instance. Such an algorithm is called a *pseudo-polynomial time algorithm*. For example, consider the *Subset Sum* problem: Given a set $S = \{s_1, s_2, \ldots, s_n\}$ of integers and an additional integer B, is there a subset S' of S such that the sum of all the integers in S' is equal to B? It can be assumed without loss of generality that B is the largest integer among the input values. (Integers in S which are larger than B cannot be part of S'.) The *Subset Sum* problem can be solved in $O(nB)$ time using dynamic programming [54]. This is a pseudo-polynomial time algorithm for the *Subset Sum* problem. On the other hand, for problems such as *Minimum Vertex Cover* and *Maximum Independent Set*, there is no pseudo-polynomial algorithm, unless $P = NP$ [54].

3.5 Formal Methods for Verification

The two main purposes of a high-level model of hardware designs are modeling and verification. To this end, various high-level hardware description languages have been proposed in the EDA industry. Some examples of such languages are SystemC, Bluespec System Verilog, and Esterel. In most cases, a high-level description of a hardware design is usually verified using traditional dynamic validation techniques. These techniques can help in removing lots of bugs present in a design. However, dynamic validation techniques have the disadvantage that even after applying these techniques, full correctness of a design cannot be guaranteed. The high-level model might still contain some bugs and this can result in the introduction of those bugs in the generated RTL description. Full correctness of a design can be ensured by using formal methods for its verification. Use of such methods for verifying high-level model of a hardware design can aid in removing all bugs from a design early in the design cycle.

Various formal verification techniques can be applied to verify such high-level models. Some of these techniques are the following:

1. *Assertion-Based Validation* – In assertion-based validation, desired properties of a design are written in a formal language. During the simulation of the design, the properties are checked to ensure their correctness.
2. *Explicit-State Model Checking* – In explicit-state model checking, various states of the design are searched exhaustively and each state is monitored to check

for the desired properties. In this technique, while traversing various states all possible paths (based on non-deterministic choices) are taken into account.

3. *Symbolic Simulation* – In symbolic simulation, instead of using binary values during simulation, symbols of the variables are propagated from inputs to outputs. In this way, many possible executions of a design can be considered simultaneously which helps in reducing the size of verification problem.

4. *Symbolic Model Checking* – In symbolic model checking, the state space of the design is explored symbolically. This is different from explicit-state model checking where state space of the design is searched explicitly.

3.5.1 Model Checking

Model checking can be successfully used to formally verify any finite state system. The main advantages of using this technique for hardware design verification is that it is fast, accurate, and generates an error trace in case a desired property does not hold for a design. The application of model checking for hardware verification consists of the following tasks:

1. *Modeling* – Modeling involves creating a model of design which can be passed as an input to a model checking tool. Such models are either generated automatically or created using some abstraction techniques to eliminate irrelevant details. High-level model of a hardware design can also be used as an input.

2. *Specification* – Specification involves the use of a formalism to specify the properties of the designs which need to be verified. For hardware verification, temporal logic is commonly used to specify such properties. Temporal logic is a formalism that allows us to describe and reason about propositions qualified in terms of time. In concurrent systems, it is used to define ordering of the events in time without introducing time explicitly. For example, using temporal logic, one can specify that a property which is not true in the present may eventually become true sometime in a future state of a system or that the property would always be true in all future states of the system. Due to this, it has found an important application in formal verification of hardware designs, where temporal logics can be used to state requirements of various designs.

   Based on whether time is assumed to have linear or branching structure, temporal logics are classified as:

   a. *Linear Time Temporal Logic* – Using linear time temporal logic (LTL), it is possible to define properties for all reachable states in a single computational path. LTL consists of a linear time structure and various temporal operators to reason about the behavior along the path.

   b. *Computational Tree Logic* – Computational Tree Logic (CTL) consists of a branching tree structure and consists of temporal operators to reason about behaviors among multiple paths. These paths are like the different branches of a tree, any of which can be taken in the actual implementation of the system.

There are many advantages of using temporal logic to specify the properties of systems. One of the main advantages is that it has a well-defined semantics and expressive power. For these reasons, temporal logic is quite commonly used for verification purposes.

3. *Verification* – Verification involves the analysis of verification results which can lead to modification of the design in case of an error. Model checking-based verification can produce an error-trace to point out the path which led to the violation of a desired property. This aids in quickly finding the problem in a design. After modification, model checking is applied again in order to ensure the correctness of the modified design.

Chapter 4
Low-Power Problem Formalization

Hardware implementations generated from CAOS can exploit the parallelism germane in the computation and execute maximal set of actions concurrently in order to reduce the latency of the design. The concurrent execution of actions may result in high peak power and dynamic power consumption in the hardware generated from such specifications. Peak power becomes an issue if large number of actions are executed in the same clock cycle. This is undesirable due to packaging, cooling, and reliability considerations. Furthermore, maximal set of actions executing in each clock cycle implies that more hardware units are working in parallel and result in high switching activity in such designs. This leads to high dynamic power consumption which increases the heat dissipation of the system and limits its battery life. Therefore, in designs generated from CAOS, strategies targeting the reduction of these power consuming activities are required [94]. In this chapter, we formulate the problems of power-optimal synthesis from CAOS. We discuss peak power and dynamic power optimization problems related to CAOS and show that both these problems are NP-hard.

4.1 Definitions

Consider a design described in terms of CAOS. Let S denote the set of its state elements. Each state element $s_i \in S$ can be assigned a value depending on its type, and a set of values of these state elements denote the state s of the design. State elements can be of different types such as registers, FIFOs, memories. A state element s_i is updated in a clock cycle if one or more of the actions updating it are executed in that cycle.

Definition 4.1 A *guard* $g(s)$ of an action is a function which maps the set of vectors, each representing the state s of the design, to set $\{0, 1\}$. For example, for the LPM module shown in Fig. 4.1, the expressions on the left side of the arrows ($\rightarrow$) in each action are the guards of the respective actions of the design. For *Action 2 (Complete)*, expression $(isLeaf(d = RAM.readResult()))$ is its guard. It reads a memory response d and checks if the lookup is complete. If the guard evaluates to *True*, then the action executes, otherwise not.

G. Singh, S.K. Shukla, *Low Power Hardware Synthesis from Concurrent Action-Oriented Specifications*, DOI 10.1007/978-1-4419-6481-6_4,
© Springer Science+Business Media, LLC 2010

Action 1 (Input):
(*true*) → {
    IPaddr = fifo1.deq();
    token = compBuffer.getToken();
    fifo2.enq(token,IPaddr[15:0],state0);
    RAM.readRequest(baseAddr + IPaddr[31:16]); }

Action 2 (Complete):
(isLeaf(d = RAM.readResult())) → {
    (token, IPaddr, s) = fifo2.deq();
    compBuf.done(token, d);
    RAM.readAck(); }

Action 3 (Circulate):
(!(isLeaf(d = RAM.readResult()))) → {
    (token, IPaddr, s) = fifo2.deq();
    fifo2.enq(token, IPaddr, s+1);
    RAM.readRequest(computeAddr(d,s,IPaddr));
    RAM.readAck();}

Fig. 4.1 LPM module design using concurrent actions

Definition 4.2 A *body* $b(s)$ of an action is a set of assignments, conditionals, or other statements which update various state elements $s \subseteq S$ of the design. If the guard of an action evaluates to *True*, then the body of the action is executed. In Fig. 4.1, the parts of various actions on the right side of the arrows ($\rightarrow$) are the bodies of these actions. For example, the body for *Action 2 (Complete)* is

$$\{ \; (token, IPaddr, s) = fifo2.deq();$$
$$compBuf.done(token, d);$$
$$RAM.readAck(); \; \}$$

Definition 4.3 An *action* a of a design is defined as a 2-tuple $\langle g(s), b(s) \rangle$, where $g(s)$ is the guard of the action and $b(s)$ is the body of the action. In the CAOS-based synthesis flow, the behavioral description of a design is given in terms of various state elements and actions of the design. For example, a *Circular pipeline* form of the LPM module design can be described using three concurrent actions (corresponding to *Input, Complete*, and *Circulate* stages) as shown in Fig. 4.1.

Definition 4.4 The *weight* w of an action is the measure of its computational effort. It can be estimated in terms of the number of operations involved in the action. Higher the number of operations in an action, larger is its weight. Since higher computational effort usually implies more power consumption, w can be used to represent the power consumed by the combinational logic corresponding to the body of an action. The body of *Action 1 (Input)* of the LPM module, shown in Fig. 4.1, involves four statements. Thus, we can assign a *weight* of four to this action. Similarly, *Action 2 (Complete)* and *Action 3 (Circulate)* can be assigned weights of three and four, respectively. Thus, $w_1 = 4$, $w_2 = 3$, and $w_3 = 4$ for the LPM module design.

Definition 4.5 An action a_i is said to have a *dependency* on another action a_j if a_j updates a state element $s \in S$ which is accessed by a_i. Let $d_{i,j}$ denote a dependency between any two actions a_i and a_j; that is, $d_{i,j}$ denotes that an action a_i is dependent on action a_j. For example, in the LPM module, *Action 2 (Complete)* and *Action 3 (Circulate)* cannot execute unless *Action 1 (Input)* launches a memory request. Thus, we can say that *Action 2* and *Action 3* are dependent on *Action 1*.

Definition 4.6 The *value* v_i of an action a_i is the measure of the priority associated with the action. Such a priority can be used to decide which actions to execute when two or more actions are enabled in a single clock cycle. A value can be assigned to an action in various ways depending on the design. One such way is to give a higher value to an action on which a large number of other actions are dependent.

Definition 4.7 Two actions a_i and a_j are said to *conflict* with each other if a_i and a_j cannot be allowed to execute concurrently in the same clock cycle. A simple example of a conflict is two actions updating the same state element. In Fig. 4.1, both *Action 1 (Input)* and *Action 3 (Circulate)* enqueue in *fifo2*. Thus, they are said to be conflicting with each other.

Definition 4.8 Two guards are said to be *mutually exclusive* if their conjunction can never evaluate to *True*. For example, in the LPM module example, guards of *Action 2 (Complete)* and *Action 3 (Circulate)* are mutually exclusive, since one is the negation of the other, and hence, both can never evaluate to *True* in the same clock cycle.

Definition 4.9 *Synchronization Points* are the points in time during the execution of a design where the values of various state variables should be the same for various possible hardware implementations of that design. These points are defined by the users of the synthesis algorithms and they indicate the points at which the synthesized design will be equivalent to the "reference design."

For the rest of the chapter we assume the following: Let A be the set of all the *actions* of a design. Let w_i be the weight of an action $a_i \in A$. Let D be the set of all dependency relationships of a design. Let SP be the set of various pre-defined synchronization points of a design. We also define the following sets:

$$g_i = \{x : \text{value of } x \in S \text{ is accessed in the guard of } a_i.\}$$
$$b_i = \{x : x \in S \text{ is accessed in the body of } a_i.\}$$
$$c_i = \{x : x \in S \text{ is updated in the body of } a_i.\}$$

In CAOS, actions describing a design may have dependency relationships among themselves. The relationships among the actions can be broadly classified into two categories:

1. *Conflict-Free Actions* – Two actions are said to be conflict free if none of the actions updates the state elements accessed (including the state elements accessed in the guards) by the other action [62]; that is, two actions a_i and a_j are

said to be conflict free if $(g_i \cup b_i) \cap c_j = \phi$ and $(g_j \cup b_j) \cap c_i = \phi$. A set of actions where all the pairs of actions are conflict free is confluent [11]; that is, a set $\alpha \subseteq A$ is confluent iff $\forall a_i, a_j \in \alpha$, a_i and a_j are conflict free.

For a confluent set of actions, re-scheduling of the actions in order to achieve the power savings is relatively simple because such actions can be scheduled in any order without altering the end result. In other words, all the possible orders of the execution of such actions will result in the same final state of the design. However, presence of such a conflict-free set of actions is not too common in realistic hardware designs.

2. *Dependent Actions* – As mentioned in Definition 5, two actions are said to have a dependency relationship if one of the actions updates state elements accessed by the other action. An action a_i is dependent on action a_j if $\exists\, x \in (g_i \cup b_i)$ such that $x \in c_j$. Such a dependency can be denoted as $d_{i,j} \in D$. A dependency graph for a set A can be represented as $G_D = (V, E)$, where $V = A$ and $E = \{(a_i, a_j) : d_{i,j} \in D\}$.

Dependent actions can be re-scheduled for power savings as long as the dependency constraints among these actions are satisfied. Such dependency among actions of a design is very common in realistic designs.

4.2 Other Details

4.2.1 Schedule of a Design

Let $S = \{s_1, s_2, \ldots, s_k\}$ be the set of k state elements of a design whose CAOS is given by an action set $A = \{a_1, a_2, \ldots, a_n\}$ containing n actions.

Let $d(s_i)$ denote the domain of element $s_i \in S$. Usually $d(s_i)$ is the Boolean domain but for a higher abstraction level of a specification, domains such as 32-bit integer, n-bit bit vector can be used.

Let $\hat{s} = <s_1, s_2, \ldots, s_k>$ be the vector of the state elements of the design. Let $\sigma(\hat{s}) = <d_1, d_2, \ldots, d_k> \in \prod_{i=1}^{k} d(s_i)$ denote the state of the design at some point.

A schedule for the design generated by the synthesis process is of the form $\{A_1 A_2, \ldots, A_n, \ldots\}$, where $A_i \subseteq A$ is a subset of actions of the design executing in the clock cycle i. A schedule may be finite or infinite depending on the transformational or the reactive nature of a design. Each $A_i \subseteq A$ of a feasible schedule should have the following property:

If $A_i = \{a_{i1}, a_{i2}, \ldots, a_{im}\}$, then

1. $\forall\, a_{ij}, a_{ik} \in A_i$, a_{ij} and a_{ik} do not conflict.
2. if the concurrent execution of the actions in A_i transforms the design from $\sigma(\hat{s})$ to $\sigma'(\hat{s})$, then $\exists$ a permutation $\prod(A_i) = \{a_{i1}^{\prod}, a_{i2}^{\prod}, \ldots, a_{im}^{\prod}\}$ of the actions in A_i such that the sequential execution of the actions in the permutation $\prod(A_i)$ also transforms the design from a state $\sigma(\hat{s})$ to $\sigma'(\hat{s})$.

4.2.2 Re-scheduling of Actions

Given an original schedule $\{A_1, A_2, \ldots, A_n, \ldots\}$ and a set of synchronization points $SP \subseteq \{1, 2, \ldots, n, \ldots\}$, let $\sigma_0(\hat{s})$ be the initial state of the design when the schedule starts executing.

A re-scheduling of the action set A generates a new schedule of the form $\{A_1', A_2', \ldots, A_q', \ldots\}$, where all the properties of a schedule are satisfied. Moreover, the re-scheduling problem demands that there exists a set $SP' \subseteq \{1, 2, \ldots, q, \ldots\}$ such that

1. $\forall\, i \in SP$, $\exists\, j \in SP'$ under the condition that starting the execution of both the schedules at $\sigma_0(\hat{s})$, $\sigma_i(\hat{s}) = \sigma_j'(\hat{s})$, where $\sigma_i(\hat{s})$ is the state after $A_1, A_2, \ldots, A_i$ sequence of the original schedule is executed and $\sigma_j'(\hat{s})$ is the state after $A_1', A_2', \ldots, A_j'$ sequence of the new schedule is executed.
2. if $(i, j) \in SM$ and $(\text{next}(i), j') \in SM$ then $j' > j$, where $\text{next}(i)$ denotes the next integer after i in the set SP and $SM \subseteq SP \times SP'$ is the synchronization map.

If a schedule β is re-scheduled to β' and there exists a set of synchronization points for which β' satisfies the above properties, then we say that the two schedules are functionally equivalent.

4.2.3 Cost of a Schedule

Given a schedule $\beta = \{A_1, A_2, \ldots, A_n, \ldots\}$, one can define the cost of a schedule in many different ways as follows:

1. The peak power can be given as

$$P_{\text{peak}}(\beta) = \max_j \left\{ \sum_{a_{ji} \in A_j,\ A_j \in \beta} w(a_{ji}) \right\} \qquad (4.1)$$

where $w(a_{ji})$ is the weight of the action a_{ji}. Thus, the peak power of a design is related to the maximum number of actions executing in any clock cycle.
2. The switching power of the schedule β can be given as

$$P_{\text{switch}}(\beta) = \sum_i P_{i,i+1} \qquad (4.2)$$

where $P_{i,i+1}$ is the energy expended in switching the inputs of the re-used functional units when the schedule moves from A_i to A_{i+1}. The switching power of a design is usually a measure of the total energy expended during switching by the design. The schedule in which the actions execute will affect the amount of switching activity of the design, thus determining its switching power.

4.2.4 Low-Power Goal

Given a schedule β of a design, one might be interested in creating a new schedule β' such that $P_{\text{peak}}(\beta') < P_{\text{peak}}(\beta)$ and/or $P_{\text{switch}}(\beta') < P_{\text{switch}}(\beta)$ to reduce the power consumption of the design. Obviously, in the absence of any additional constraints, it is easier to reduce the peak power by just scheduling one action of the design in each clock cycle. But such a schedule might not be acceptable due to the latency considerations. Usually for a design, maximum peak power is specified as a constraint. So in each clock cycle maximum possible set of actions can be executed under the given peak power constraint in order to minimize the degradation in the latency of the design.

In other cases, constraint on the latency of the design can be specified. For example, for each $(i, j) \in SM$ a constraint on the latency can be specified as

$$\max_{(i,j)\,\in\, SM} \frac{\text{next}(j) - j}{\text{next}(i) - i} < \mu \tag{4.3}$$

where μ is a pre-defined constant denoting the maximum latency degradation factor for the re-scheduling. Let $P_{\text{avg}}(\beta)$ be the average peak power of the schedule β. Using the given constraint on the latency of the design, average peak power constraint of the new schedule can be estimated as

$$P_{\text{avg}}(\beta') = \frac{P_{\text{avg}}(\beta)}{\mu} \tag{4.4}$$

This constraint can be used to meet the average peak power goal of the re-scheduled design.

4.2.5 Factorizing an Action

Let a_i be an action updating multiple state elements of a design during its execution; that is, c_i contains multiple state elements. Therefore, a_i can be split into a set of sub-actions $a_i^1, a_i^2, \ldots, a_i^z$. We call such a split "factorizing an action." A strategy which schedules such parts (sub-actions) of an action for power reduction is explained in Chapter 5.

Since each action executes atomically in CAOS, the scheduling of such factorized parts of an action a_i should be done such that no other action a_j dependent on a_i is scheduled before the parts of a_i on which a_j is dependent have been executed. So if $(g_j \cup b_j) \cap c_i$ and a_i is factorized, then a_j should not be scheduled until all the parts of a_i on which a_j is dependent have been executed.

Moreover, a part a_i^t of an action a_i should be executed in the present clock cycle if an action a_j executing in the same clock cycle updates the state accessed by a_i^t.

4.3 Formalization of Low-Power Problems

4.3.1 Peak Power Problem

In hardware synthesis from CAOS, a control circuit selects the actions which can be executed in a clock cycle. The number of actions that execute in a particular clock cycle determines the peak power consumed in that cycle. Higher peak power is undesirable due to packaging, cooling, and reliability considerations, thus making peak power minimization an essential part of the low-power synthesis goal.

For the "reference design," let G be the maximal set of actions which execute in a particular clock cycle. These will be the actions which have their guards evaluate to *True* and do not conflict with each other. If two actions, having their guards evaluate to *True*, are conflicting with each other, then the one with the higher user-defined urgency is chosen to be in G; in case the two actions have the same urgency, one is chosen arbitrarily.

Let f_i denote a variable which takes a value 1 if the action a_i should be executed in a particular clock cycle to achieve the low peak power goal, and 0 otherwise. Now, if P_{peak} is the maximum allowable peak power for the design, then for each cycle, the peak power minimization problem can be formalized as

$$\text{maximize} \left(\sum_{i \in G} v_i * f_i \right) \tag{4.5}$$

under the following constraints:

1. $\sum_{i \in G} (w_i * f_i) \leq P_{peak}$.
2. Satisfy all dependencies $d_{i,j} \in D$ between various actions of the design.
3. The new schedule of the execution of actions should produce behavior which is equivalent to the schedule of the "reference design" at all pre-defined synchronization points $sp \in SP$.

4.3.2 Dynamic Power Problem

As mentioned earlier, dynamic power is a major component of the total power consumption of a design. Consequently, power optimization techniques targeting the reduction of the dynamic power form an integral part of low-power design strategies. Dynamic power depends on the switching activity in the circuit. Large amount of switching at the inputs of various functional units of the design leads to high dynamic power consumption. Thus, dynamic power can be significantly reduced by reducing the amount of switching that occurs at the inputs of the functional units between different clock cycles.

Let P_{switch} denote the total switching power consumed by the design. Now, the switching power minimization problem can be formalized as

$$\text{minimize } (P_{\text{switch}}) \tag{4.6}$$

under the following constraints:

1. Satisfy all dependencies $d_{i,j} \in D$ between various actions of the design.
2. The new schedule of the execution of actions should produce behavior which is equivalent to the schedule of the "reference design" at all pre-defined synchronization points $sp \in SP$.

4.3.3 Peak Power Problem Is NP-Complete

Here, we sketch the proof of NP-completeness of the peak power optimization problem by showing the reduction from the 0/1 *Knapsack* problem. The 0/1 *Knapsack* problem [54] states that given a set of items, each with a cost and a value, determine if an item should be included in a collection so that the total cost is less than some given cost and the total value is as large as possible. The number of each item is restricted to zero or one. It is known that the 0/1 *Knapsack* problem is NP-complete.

The 0/1 *Knapsack* problem can be reduced to the peak power optimization problem as follows. The items in the *Knapsack* problem can be thought of as the actions of the peak power problem. The cost of each item in the *Knapsack* problem can correspond to the weight of each action. The total cost can be thought of as the peak power P_{peak}. Each action also has an associated value. Thus, if a solution to the 0/1 *Knapsack* problem exists, then the peak power optimization problem can also be solved, and vice versa. Hence, it can be claimed that a solution to the 0/1 *Knapsack* problem exists if and only if there exists a solution of the peak power optimization problem.

Since the peak power optimization problem clearly belongs to NP, and the 0/1 *Knapsack* problem (which is NP-complete) can be reduced to the peak power optimization problem, this implies that the peak power optimization problem is also NP-complete [54]. Thus, no polynomial time algorithm exists for the peak power minimization problem mentioned above, which justifies the use of heuristic algorithms for the peak power minimization problem.

4.3.4 Dynamic Power Problem Is NP-Complete

Now, in order to analyze the complexity of the dynamic power optimization problem, consider the problem of *Directed Hamiltonian Path*, which is stated as follows: Given a directed graph $G = (V, E)$ and a function $W : E \to \mathbb{R}$ which maps each edge in the set E to a real number, find a path that includes every vertex of set V exactly once; that is, find a path $p = (v_1, v_2, \ldots, v_n)$ containing all the vertices in V such that

$$\forall\, v_i \in V, v_j \in V, i \neq j, v_i \neq v_j \text{ and}$$

$$\sum_{k=1}^{n} W(v_k, v_{k+1}) \text{ is minimized}$$

The *Directed Hamiltonian Path* problem is NP-complete [54].

Now, consider a special case of the dynamic power optimization problem. Let A' be a set of actions, f be a functional unit, and S be a function which relates a pair of actions to the amount of switching power p consumed due to the switching at the inputs of f when both the actions of the pair use f in consecutive clock cycles; that is, $S(a_i, a_j) = p$, where $a_i \in A', a_j \in A', a_i \neq a_j$. Given A', f, and S such that each action belonging to A' uses f to compute an operation, the problem is to find a schedule that includes every action of the set A' exactly once such that minimum switching power is consumed.

The *Directed Hamiltonian Path* problem can be reduced to the special case of the dynamic power optimization problem. For this, the set of vertices V of the *Directed Hamiltonian Path* problem can be replaced with the set of actions A' of the dynamic power optimization problem, and the function W can be replaced with the function S. This reduction implies that a solution to the *Directed Hamiltonian Path* problem exists if and only if there exists a solution to the dynamic power optimization problem mentioned above. Thus, the dynamic power optimization problem is NP-hard. Also, the dynamic power optimization problem, in general, belongs to NP. Hence, it can be concluded that the dynamic power optimization problem is NP-complete [54].

Chapter 5
Heuristics for Power Savings

As discussed in Chapter 4, the optimization problems germane in the low-power CAOS-based synthesis process are NP-hard. Thus, heuristics are needed to solve these problems efficiently. In this chapter, we discuss a class of heuristics targeting the reduction of peak power and average dynamic power in the scheduling and allocation phases of the CAOS-based synthesis process. We show some numerical examples illustrating the use of such heuristics during CAOS-based synthesis.

The proposed approaches are based on first creating a schedule based on highest possible parallelism among the actions of a design by analyzing the conflict graph as done in [10]. The heuristics create alternative schedules which will reduce power by reducing the parallelism in some cases, thereby reducing the peak power, and in some cases by cleverly allowing resource sharing such that the switching power can be reduced. The figures of merit for the heuristics we provide are measured in terms of power savings with respect to the most concurrent schedules. One important aspect of the re-scheduling approach is that the actions that are re-scheduled may lead to an implementation of the hardware that is not clock cycle equivalent to the original one, except at certain pre-defined synchronization points. Such equivalences are quite common in hardware design when re-timing and other optimizations are done. In other words, hardware generated by these heuristics is functionally equivalent to the one that does not consider power optimizations and allows maximal parallelism. We call the hardware implementation that uses maximal concurrency the "reference design". This might also mean that the latency of certain outputs in the synthesized result may suffer on using the low-power optimizations, and the goal is to minimize such penalties in latency. In most cases, however, the proposed optimizations will reduce area, thereby providing solutions which are better in at least two design parameters.

Section 5.1 presents simple heuristics for the minimization of peak power and dynamic power in designs generated from CAOS. The proposed heuristics demonstrate how scheduling techniques can be exploited for power reductions during CAOS-based synthesis process. These techniques can be further refined as per various latency and synchronization constraints. Section 5.2 discusses various such refinements.

G. Singh, S.K. Shukla, *Low Power Hardware Synthesis from Concurrent Action-Oriented Specifications*, DOI 10.1007/978-1-4419-6481-6_5,
© Springer Science+Business Media, LLC 2010

5.1 Basic Heuristics

5.1.1 Peak Power Reduction

In the CAOS-based synthesis approach, two non-conflicting actions can be executed concurrently in the same clock cycle provided their concurrent execution produces a behavior that corresponds to at least one sequential ordering of the execution of those actions [62]. Thus, two non-conflicting actions a_i and a_j can be executed concurrently if their concurrent execution produces a behavior which corresponds to either the execution of a_i after a_j, or the execution of a_j after a_i. The peak power minimization heuristic, Algorithm 1, exploits this property of the CAOS-based synthesis approach to minimize the peak power of a design. Figure 5.1 gives a list of variables used to describe Algorithm 1.

A : Set of actions.
S : Set of state elements of the design.
T : Table containing all the possible pairs of non-conflicting actions of a design
    and their associated orderings.
G : Set of all the non-conflicting actions whose guards evaluate to *True* in a
    clock cycle.
R : Set representing an ordering of actions in G.
W : Set of actions which should be executed in a clock cycle in order to limit
    the peak power.
P_{peak} : Maximum allowable peak power.
P_W: Power consumed in a clock cycle.
w_i : Estimated weight of an action a_i.

Fig. 5.1 List of the variables used in Algorithm 1

5.1.1.1 Description of Algorithm 1

Algorithm 1 starts by statically constructing a table T containing all the possible pairs of non-conflicting actions and their associated orderings. The behavior of each ordering corresponds to the concurrent execution of the actions of the pair with which the ordering is associated.

For each clock cycle during the execution of the design, Algorithm 1 returns a set of actions W that should be executed in that clock cycle in order to keep the maximum peak power of the design within limits. P_W denotes the maximum power consumed in each cycle. For each clock cycle, initially W is empty and P_W is zero. Using the table T and the set G of two or more non-conflicting actions whose guards have evaluated to be *True* in a particular clock cycle, the heuristic dynamically determines an ordering (whose behavior corresponds to the concurrent execution of the actions in the set) of the actions of set G. This ordering of the actions is stored in set R (lines 10–14).

For each cycle, starting from the first element of the set R and following the ordering in R, the algorithm keeps moving actions from set R to set W as long as

Algorithm 1 – Heuristic for Peak Power Minimization

1. Construct the table T containing all the possible pairs of non-conflicting actions of a design and their associated orderings.
2. Let G be the set of all the non-conflicting actions whose guards evaluate to *True* in a clock cycle.
3. Let R be the set representing an ordering of actions in G. Initially, $R = NULL$;
4. Let W be the set of actions which should be executed in the present clock cycle in order to keep the peak power within limits.
5. Let P_W be the sum of weights of all the actions in W.
6. For each clock cycle
7. {
8. $W = NULL$;
9. $P_W = 0$;
10.   If $(R == NULL)$
11.   {
12.   Compute G;
13.   Construct R using G and T;
14.   }
15.   While $((P_W < P_{peak})$ AND $(R \neq NULL))$
16.   {
17.   Pick the first action, a_i from R;
18.   If $(P_W + w_i \leq P_{peak})$
19.   {
20. $P_W = P_W + w_i$;
21.   Add a_i to W;
22.   Remove a_i from R;
23.   }
24.   }
25.   In the present clock cycle, execute all actions belonging to W;
26. }

the peak power remains less than or equal to P_{peak} (lines 15–24). If the set R is still not empty when the peak power exceeds P_{peak}, then the present set R (with remaining actions) is used in the next clock cycle to determine which actions from the set should be executed in the next cycle. For all the future clock cycles, this same set R (with remaining actions) is used until R becomes empty in some future clock cycle. At that point, in the clock cycle after the cycle in which R becomes empty, set R is reconstructed using the table T and the new set G which will now contain all the non-conflicting actions whose guards have evaluated to be *True* in that clock cycle.

Thus, the heuristic makes sure that the peak power for each clock cycle does not exceed P_{peak}. Moreover, since the actions to be executed in a clock cycle are selected on the basis of the orderings in the set R, therefore for the schedule of actions generated using this algorithm, the starting points of all the clock cycles where a new set R is reconstructed using set G and table T will represent the synchronization points of the schedule. Thus, the behavior of the schedule obtained by applying this heuristic remains equivalent to the schedule of the "reference design" at these synchronization points.

5.1.2 Dynamic Power Reduction

While describing a design using CAOS, dynamic power reduction can be achieved by selecting a schedule in which the switching activities at the inputs of various functional units are reduced. Algorithm 2 illustrates a heuristic which generates a schedule that aids in the reduction of the switching of the inputs.

Let O denote the set of all the operators that can be operated on the elements of set S. Let F be the set of functional units available for implementing the design. Each operator of the set O is involved either in an action a_i or not. Thus, a subset of O represents the operators used in the action a_i. Each action a_i accesses a set of state elements $S_i \subseteq S$. Let $VARS$ be the function which maps the set of actions A to various such subsets of set S; that is, $VARS : A \rightarrow 2^S$. For example, assume that an action a_i requires an operation $l = (m + n)$ to occur when its executed, then the state elements m and n are accessed when this action is executed, and so they belong to S_i.

Let $VAROPS$ be the function that maps a pair (a_i, s_i), where $a_i \in A$, $s_i \in S$ to a subset of O.

Let $VARFUN$ be the function that maps a pair (a_i, s_i), where $a_i \in A$, $s_i \in S$ to a subset of F.

Figure 5.2 gives a list of sets and functions used to describe Algorithm 2.

F : *Set of functional units.*
O : *Set of operators.*
$VARS : A \rightarrow 2^S$
$VAROPS : (A \times S) \rightarrow 2^O$
$VARFUN : (A \times S) \rightarrow 2^F$

Fig. 5.2 List of the sets/functions used in Algorithm 2

5.1.2.1 Description of Algorithm 2

Algorithm 2 utilizes the fact that if a state element is involved in the same operation in two actions a_i and a_j executing in different clock cycles, then the switching activity can be reduced by sharing the same functional unit for implementing that particular operation of the two actions. Thus, the algorithm first identifies (lines 4–17) all such pairs of actions and then forces them to execute in different clock cycles (line 28) to enable maximum sharing of functional units among them. It starts with an empty set P which contains elements of the form (a_i, a_j, k), where a_i and a_j denote the actions that involve one or more same state elements in k number of common operations. It then picks (line 4) an arbitrary action a_i from the set A and checks (lines 6–16) if there exists another action a_j in the set A, such that a_i and a_j involve one or more same state elements in some common operation. In Algorithm 2, the number of such common operations is denoted by variable *commonOp*. If such an action a_j exists in A, then an element $(a_i, a_j, commonOp)$ is added to P as shown in lines 14 and 15 of Algorithm 2.

Algorithm 2 – Heuristic for Switching Power Minimization

1. Let P denoted the set containing elements of the form (a_i, a_j, k), where a_i and a_j are the actions that involve one or more same state elements in k number of common operations.
2. Let *commonOp* be the variable denoting the number of common operations involving same state element between any two actions.
3. Initially, $P = NULL$ and *commonOp* $= 0$;

4. For $i = 1$ *to* $|A|$, where $|A|$ denotes the number of elements of set A.
5. {
6.   For $j = (i + 1)$ *to* $|A|$
7.   {
8.   *commonOp* $= 0$;
9.   For each state element $x \in (VARS(a_i) \cap VARS(a_j))$
10.   {
11.   if $((VAROPS(a_i, x) \cap VAROPS(a_j, x)) \neq NULL)$
12.   *commonOp* $=$ *commonOp* $+ |(VAROPS(a_i, x) \cap VAROPS(a_j, x))|$;
13.   }
14.   if $(commonOp > 0)$
15.   Add $(a_i, a_j, commonOp)$ to the set P.
16.   }
17. }

18. Arrange elements of the set P in decreasing order of k.
19. Choose the first element of the set P and assign it to p_i.
20. While p_i is not the last element of P
21. {
22.   For a_i and a_j corresponding to the element p_i
23.   Remove all the entries of P, other than p_i, which contains either a_i or a_j.
24.   Choose the next element of P.
25. }

26. For each element p_i of P
27. {
28.   If actions a_i and a_j corresponding to p_i do not conflict and do not have mutually exclusive guards, then define a conflict between a_i and a_j so that they never execute in the same cycle, thus allowing functional units sharing.
29.   For each state element $x \in (VARS(a_i) \cap VARS(a_j))$
30.   {
31.   For each operator $y \in (VAROPS(a_i, x) \cap VAROPS(a_j, x))$
32.   Allocate same functional unit $f \in (VARFUN(a_i, x) \cap VARFUN(a_j, x))$ to y;
33.   }
34. }

In order to achieve switching power savings, only those elements of set P will be useful which correspond to maximum number of common operations k between actions a_i and a_j. Thus, in lines 18–25, Algorithm 2 removes all the other elements from set P.

At this step in the algorithm (line 26), actions in each element $(a_i, a_j, commonOp)$ of the set P have the relationship that if the actions a_i and a_j belonging to an element of P do not execute in the same clock cycle, then some of the functional units used to implement the operators belonging to the first action a_i of the element can be shared in another clock cycle to implement the similar operators in the second action a_j of the element (these will be the operators in a_j with one or more same inputs as the operators of a_i). This will result in the reduction of the switching activity across those functional units because the input of the functional units will not change frequently. Thus, power savings can be achieved if the actions of each pair in P do not execute in the same clock cycle. And as shown in the algorithm, this can be achieved by assigning a conflict between the actions of each element of P. Therefore, in the later part (lines 26–34) of the algorithm, each element of set P is considered and a conflict is defined between the corresponding actions of that element.

The time complexity of the heuristic Algorithm 2 is of the order $|A|^2$, where $|A|$ denotes the number of actions of the design. This quadratic dependence of the time complexity on the number of actions is due to the fact that each action is compared with all the other actions to check if there exists a potential for switching power savings by executing any two actions in different clock cycles.

5.1.3 Example Applications

5.1.3.1 Peak Power Reduction

On applying the peak power minimization heuristic, Algorithm 1 to various designs like Crossbar Switch, FIR Filter, LPM module, we achieved the required low peak power design goal. As an illustration of the performance of Algorithm 1, we consider the LPM module design of Fig. 4.1. In that design, the body of *Action 1* consists of four operations, and so we can assign a weight w_1 of 4 to this action. As mentioned earlier, w_1 denotes the computational effort of *Action 1* and hence it corresponds to the power consumed during the execution of this action. Similarly, since the bodies of *Action 2 (Complete)* and *Action 3 (Circulate)* consist of three and four operations, respectively, we can assign weights $w_2 = 3$ and $w_3 = 4$ to these actions.

As shown in Fig. 4.1, the *Action 1 (Input)* and the *Action 3 (Circulate)* actions of the design enqueue in FIFO *fifo2*. Since these actions update the same state element *fifo2*, these two actions are said to be conflicting with each other and hence cannot execute simultaneously in the same clock cycle. We assign a higher urgency to *Action 3* which would mean that whenever the guards of these two actions evaluate to *True*, then in order to avoid any deadlocks, *Action 3* will be executed.

Now, notice that for a given clock cycle, the guards of *Action 2 (Complete)* and *Action 3 (Circulate)* cannot evaluate to be *True* because they are mutually exclusive. This means that *Action 2* and *Action 3* will never execute in the same clock cycle. Thus, *Action 3 (Circulate)* will not execute with any of the other two actions in a given clock cycle. However, *Action 1 (Input)* and *Action 2 (Complete)* can execute in the same clock cycle if their guards evaluate to *True*. Assuming that there is no constraint on the peak power consumed by the design, Table 5.1 lists all the possible combinations of actions of the LPM design that can execute in the same clock cycle and the corresponding estimated power consumption (based on the weights of the actions) for that clock cycle. For example, when actions *Action 1 (Input)* and *Action*

Table 5.1 Combinations of the execution of actions and the associated power consumption

Possible combinations of actions of the LPM design that can execute in the same clock cycle	Estimated power consumption in that clock cycle units
Input	4
Complete	3
Circulate	4
Input, complete	7

2 (Complete) execute in the same clock cycle, then the power consumed can be estimated to be equal to $(w_1 + w_3) = 4 + 3 = 7$ units.

Lets assume that the maximum allowable peak power P_{peak} for the LPM design is 5 units. In that case, among all the possible combinations of the actions that can execute in the same clock cycle (Table 5.1), only the combination corresponding to concurrent execution of *Action 1 (Input)* and *Action 2 (Complete)* in a clock cycle is undesirable. This is because if these two actions execute in the same clock cycle, then the power consumed for that cycle will be 7 units. For all the other combinations listed in Table 5.1, the peak power consumed in a clock cycle is below 5 units.

On applying Algorithm 1 to the LPM design, it first generates a table containing all the possible pairs of non-conflicting actions of a design and their associated orderings. Since *Action 1 (Input)* and *Action 3 (Circulate)* conflict with each other, thus the table will just have two entries, the first corresponding to the pair containing *Action 1 (Input)* and *Action 2 (Complete)*, and the second corresponding to the pair containing *Action 2 (Complete)* and *Action 3 (Circulate)*. We are only interested in the first entry of the table because the actions corresponding to the second entry have mutually exclusive guards, and hence these actions will never execute concurrently. The ordering corresponding to the first entry of the table will be {*Action 2 (Complete)*, *Action 1 (Input)*}. This is because *Action 2 (Complete)* dequeues tuple from *fifo2*, whereas *Action 1 (Input)* enqueues into *fifo2*, and hence an ordering where *Action 2 (Complete)* executes before *Action 1 (Input)* is valid.

Now, in the clock cycles when the guards of *Action 1 (Input)* and *Action 2 (Complete)* evaluate to *True*, the algorithm will only allow *Action 2 (Complete)* to execute (because it is the first action in the ordering corresponding to the first entry of the table T) and will postpone the execution of *Action 1 (Input)* to the next clock cycle in order to meet the peak power constraint for the design. In other words, Algorithm 1 does not allow actions *Action 1 (Input)* and *Action 2 (Complete)* to execute in the same clock cycle.

Table 5.2 shows all the possible combinations of the actions of the LPM design that can execute in the same clock cycle on applying Algorithm 1 with the peak power constraint, $P_{peak} = 5$ units. It also shows the corresponding estimated power consumption (based on the weights of the actions) for that clock cycle. For any given clock cycle, only one action is executed based on the heuristic. This results in a maximum peak power of 4 units for any cycle, thus meeting the maximum allowable peak power goal of 5 units. Similar results are obtained when Algorithm 1 is applied on other realistic designs described using CAOS.

Table 5.2 Combinations of the execution of actions and the associated power consumption on applying Algorithm 1 when $P_{peak} = 5$ units

Possible combinations of actions of the LPM design that can execute in the same clock cycle	Estimated power consumption in that clock cycle units
Input	4
Complete	3
Circulate	4

5.1.3.2 Dynamic Power Reduction

The switching power minimization heuristic, Algorithm 2, when applied to various designs achieved the required switching power savings. As an example, consider the design of a *Packet Processor*. Figure 5.3 shows the action-oriented description of that design. There, *Action 1 (Receive)* corresponds to the actions that the processor executes when a packet arrives. Then, actions in *Action 2 (Process)* process that packet. And finally, *Action 3 (Send)* corresponds to the departure of the packet.

Action 1 (Receive):
$true \rightarrow \{$ //Receive packet. $\wedge$
$a := a + 1; \wedge$
$c := (a + b) * 10; \}$

Action 2 (Process):
$(a > 0) \rightarrow \{$ //Process packet. $\wedge$
$b := b + 1; \wedge$
$a := a - 1; \wedge$
$c := (a + b) * 5 ; \}$

Action 3 (Send):
$(b > 0) \rightarrow \{$ //Send packet. $\wedge$
$b := b - 1;\}$

Fig. 5.3 Packet processor design

When Algorithm 2 is applied on the *Packet Processor* design, it successfully identifies various common operations involving one or more same state elements in various actions of the design. For example, the heuristic identifies that both *Action 1 (Receive)* and *Action 2 (Process)* involve the state elements a and b in an addition operation. Thus, it defines a conflict between these two actions so that they never execute in the same clock cycle. The same functional unit *ADDER2* is then allocated to the operation $(a + b)$ in both the actions. Similar allocation of other functional units is done for all the other elements of the set P generated during the application of Algorithm 2. This optimizes the switching power for the design. We applied Algorithm 2 to other realistic designs and noticed reduction in their switching power. Thus, Algorithm 2 can be used to facilitate the achievement of low-power synthesis goal.

5.2 Refinements of Above Heuristics

Algorithm 1 presented in Section 5.1.1 considers a single cycle of the original schedule at a time and throttles the parallelism to meet the peak power constraint of a
design. Algorithm 1 can be further refined by considering the actions executing over
multiple clock cycles of the original schedule, and re-scheduling these actions such
that the degradation in the latency of the design is minimized while meeting its peak
power constraint. While performing such re-scheduling of the actions over multiple
clock cycles, dependency constraints among various actions of the design need to
be satisfied to arrive at the power-efficient schedule. Factorization of the actions of
a design into smaller actions is another technique that can be used to reduce the
peak power as well as dynamic power of a design while minimizing its latency
degradation. The execution of such parts of an action will also be constrained by the
dependencies among different parts of that action as well as their dependencies with
the other actions of the design. In this section, we present various heuristics which
implements such re-scheduling and factorization-based refinements to improve the
algorithms discussed in Section 5.1.

We assume that given a design, there is a base-line (original) schedule γ which
is generated by scheduling maximal possible set of actions in each clock cycle.
Such a schedule reduces the latency of the design without any regard to the cost.
In this section, any new schedule generated from a re-scheduling strategy is always
compared against this base-line schedule γ.

5.2.1 Re-scheduling of Actions

5.2.1.1 Peak Power Reduction

Algorithm 1 discussed in Section 5.1.1 postpones the execution of some actions to
the future clock cycles to meet the peak power constraints of the design. No new
actions are considered for execution unless all the postponed actions are executed.
This increases the latency of the design. A strategy that allows the execution (in
future clock cycles) of those actions which are not dependent on the postponed
actions will improve the latency of the design. Algorithm 3 is based on such a
strategy which re-schedules the actions of a design based on their dependency relationships under the given the peak power constraint of the design. By allowing
the execution of the actions which are not dependent on the postponed actions, it
allows maximal set of actions to execute in a clock cycle as long as the peak power
constraint is satisfied.

Here again, let P_{peak} be the maximum allowable peak power of the design. Let us
assume that in some clock cycle c the peak power consumed by the design exceeds
P_{peak}. Let $A' \subset A$ be the set of all the actions that execute in clock cycle c in
the base-line schedule. Since the peak power in clock cycle c exceeds P_{peak} the
algorithm chooses a subset, say A'_1, of the actions in A' which can execute in c
without violating the peak power constraint. The actions of A'_1 are chosen on the

---

Algorithm 3 – Refined Heuristic for Peak Power Minimization

---

1. Let $A' \subset A$ be the set of all the actions that do not conflict among themselves and whose guards evaluate to *True* in the present clock cycle. If two actions conflict, one with the higher priority is chosen to execute.
2. Let A'_1 be the set of actions that are allowed to execute in the present clock cycle to meet the peak power constraint.
3. Let A'_2 be the set of actions whose execution is postponed to the future clock cycles in order to meet the peak power constraint. Initially, $A'_2 = NULL$.
4. Let P be the power consumed in the present clock cycle. Let i be a counter.
5. While ( ) {
6.   For the present clock cycle, compute A'.
7.   If $((A'_2 == NULL)$ AND $(A' == NULL))$ break;
8.   Order the elements of A' based on their dependency relationships. If a dependency $d_{i,j}$ exists between any two actions a_i and a_j belonging to A' then a_i is placed before a_j in the ordered set. Actions having no dependency among themselves can be placed in any order.
9.   If $(A'_2 \neq NULL))$ {
10. $i = 1$; Choose the ith element of A'_2 and assign it to a_i.
11.   While $(P < P_{peak})$ {
12.   If there exists no other action $a_j \in A'_2$ placed before $a_i \in A'_2$ for which $d_{i,j}$ or $d_{j,i}$ is an element of D AND a_i does not conflict with any of the actions in A'_1 {
13.   If $(P + w_i \leq P_{peak})$ {
14.   Add a_i to A'_1; Remove a_i from A'_2;
15. $P = P + w_i$;
16.   }
17.   Else $i = i + 1$;
18.   If the ith element of A'_2 exists
19.   Choose the ith element of A'_2 and assign it to a_i.
20.   Else break;
21.   }
22.   }
23.   }
24.   If $(A' \neq NULL))$ {
25. $i = 1$; Choose the ith element of A' and assign it to a_i.
26.   While $(P < P_{peak})$ {
27.   If there exists no other action $a_j \in A'$ placed before $a_i \in A'$ for which $d_{j,i}$ is an element of D AND there exists no action $a_k \in A'_2$ for which $d_{i,k}$ or $d_{k,i}$ belongs to D AND there exists no action $a_l \in A'_1$ for which $d_{i,l}$ belongs to D AND a_i does not conflict with any of the actions in A'_1 {
28.   If $(P + w_i \leq P_{peak})$ {
29.   Add a_i to A'_1; Remove a_i from A';
30. $P = P + w_i$;
31.   }
32.   Else $i = i + 1$;
33.   If the ith element of A'_2 exists
34.   Choose the ith element of A'_2 and assign it to a_i.
35.   Else break;
36.   }
37.   }
38.   }
39.   Execute all the actions in A'_1; Append all the elements of A' to A'_2 maintaining their order;
40. $A' = NULL$;
41. }

---

basis of their dependency relationships. An action a_i which accesses one or more state elements updated by another action a_j is preferred over a_j to execute in the clock cycle c; that is, if one of the elements of D is $d_{i,j}$ then action a_i is preferred over a_j. In such cases, an anti-dependency is said to exist among a_i and a_j since the reads done by a_i need to happen before the writes of a_j.

The set of remaining actions, say A'_2, are postponed to the future clock cycles; that is, in the next clock cycle all the actions belonging to A'_2 are considered for execution along with other actions whose guards evaluate to *True* in that clock cycle. Dependency relationships among all these actions are taken into account to select the actions for execution in that clock cycle. Thus, Algorithm 3 uses the sequential ordering of the actions based on their dependency relationships to re-schedule the actions in order to meet the peak power constraint of the design.

For some designs, their latency constraint is specified instead of the maximum allowable peak power constraint. In such cases, an estimation of the average peak power constraint of the design can be done using Eqs. (4.3) and (4.4) in Chapter 4. Algorithm 3 can then be used to meet the estimated peak power constraint.

Example

Let us consider a design with five actions a_1, a_2, a_3, a_4, and a_5 such that in the base-line schedule (with no peak power constraints)

1. Actions a_1, a_2, and a_3 execute in clock cycle c.
2. Actions a_4 and a_5 execute in clock cycle after c.
3. Action a_1 does not conflict with a_5.
4. Set D representing dependency relationships of the design is given as, $D = \{d_{2,1}, d_{3,2}, d_{5,3}, d_{4,2}, d_{1,4}\}$.

Figure 5.4 shows the order in which these rules are executed in the base-line schedule. Arrows in the figure denote the dependency relationships among various actions of the design. A dependency of the form $d_{i,j}$ is denoted by an arrow from action a_j to a_i. Such a dependency among any two actions executing in the same clock cycle is also known as an anti-dependency (writes after reads).

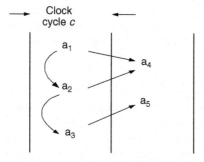

Fig. 5.4 Schedule under no peak power constraints

Suppose that the peak power constraint of the design does not allow three or more actions to execute in the same clock cycle. Let us assume that all the actions of the design consume same amount of power when executed. Thus, the power consumed due to the execution of actions a_1, a_2, and a_3 in c will exceed the maximum allowable peak power P_{peak} of the design. This means that all the three actions cannot execute concurrently in clock cycle c.

Applying Algorithm 3 on this schedule, a new schedule as shown in Fig. 5.5 is obtained which meets the peak power constraint of the design. In Algorithm 3, when the actions executing in the clock cycle c are ordered based on their dependency relationships, following ordering is obtained $\{a_3, a_2, a_1\}$. The concurrent execution of these actions is equivalent to their sequential execution in this order. Therefore, actions a_3 and a_2 are allowed to execute in clock cycle c and action a_1 is postponed to the next clock cycle to meet the peak power constraint. This is because a_1 is not dependent on actions a_2 and a_3. In Fig. 5.5, the backward arrow from a_1 to a_2 denotes the anti-dependency between the two actions (writes by a_1 should occur after reads by a_2).

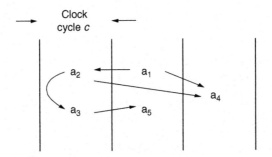

Fig. 5.5 Schedule by applying Algorithm 3 under peak power constraints

Moreover, actions a_1 and a_5 are allowed to execute concurrently in the same clock cycle (the one after clock cycle c) because they do not conflict with each other and a_5 is not dependent on a_1. Otherwise, a_5 would not have been allowed to execute concurrently with a_1 and would be executed after the execution of a_1. Also, action a_4 is dependent (data dependency) on action a_1. Therefore, the execution of action a_4 is postponed to the clock cycle after the cycle in which a_1 is executed. It can be noted that the new schedule shown in Fig. 5.5 meets the peak power constraint of the design but increases the latency of the design by one clock cycle.

5.2.1.2 Dynamic Power Reduction

The re-scheduling of the actions of a design can also be used to reduce the dynamic power of the design by re-scheduling its actions based on their dependency relationships such that the sharing of the functional units (having same inputs in the consecutive cycles) is increased in order to reduce the switching activity of the design. Algorithm 2 discussed in Section 5.1.2 is based on the strategy of reducing the dynamic power of a design by re-scheduling its actions. In that algorithm, two actions involving one or more same state elements in similar operation are never scheduled in the same clock cycle so that same functional resource can be assigned to implement those operators in order to reduce the switching activity of the design. The algorithm assumes that various dependency relationships among the actions are maintained during this process of re-scheduling. But in some cases, such re-scheduling of the actions will violate the dependency constraints among the actions. In such cases, appropriate actions need to be executed (similar to as done in Algorithm 3) in each clock cycle to satisfy various dependency constraints of the design.

5.2.2 Factorizing and Re-scheduling of Actions

5.2.2.1 Peak Power Reduction

Use of factorization of the actions of a design for peak power minimization is based on the fact that an action which is not allowed to execute in the present clock cycle due to the peak power constraints can be sub-divided into smaller parts such that some of these parts can be executed in that clock cycle. It might be possible to execute such parts of the action because they consume less power than the execution of the action itself. The remaining parts of the action can be executed in the next clock cycle. But this will result in the violation of the atomicity of the action; that is, all the parts of an action won't execute in the same clock cycle. Thus, the actions are sub-divided such that the atomicity of the factorized actions is maintained even when different parts of an action execute in different clock cycles. For this, parts of the action are executed in different clock cycles based on their dependency relationships among themselves as well as with the other actions of the design. Using this strategy, factorized actions are spread over multiple clock cycles to satisfy the peak power constraints. Such factorization of the actions of a design allows maximum possible computation to occur in each clock cycle under the given peak power constraint, thus minimizing the degradation of the latency of the design.

An action should be factorized such that there exists at least one sequential ordering of the execution of its parts whose behavior corresponds to the atomic execution of that action. This means that the parts of the actions should have a possible sequential ordering based on the dependency relationships among them. If an action is divided into smaller parts but there exists no such ordering of the parts, then that action cannot be factorized into those parts.

The factorization of the actions into lower granularity parts provides the opportunity for reducing the latency of the design while satisfying its peak power constraints. If some of the parts of the action can be executed in the clock cycle where the action itself could not be executed, then in the next clock cycle actions which are dependent on the executed parts of the factorized action can be considered for execution. Thus, factorization of the actions for peak power minimization may improve the latency of the design since more computation can take place in each clock cycle while meeting the peak power constraints of the design.

Let a_i be an action, with guard evaluated to *True* in a clock cycle c. Suppose that a_i does not conflict with any other actions executing in c. Also, let us assume that if a_i is allowed to execute in the clock cycle c then the power consumed in c will exceed the maximum allowable peak power P_{peak} of the design. If the factorization of the actions is not considered (as in Algorithm 3), then the execution of action a_i will be postponed to the future clock cycles. However, if the factorization of the actions is considered, then a_i can be divided into smaller parts. Suppose we divide action a_i into two parts a_i^1 and a_i^2. Let us assume that if a_i^2 is executed after a_i^1, then the behavior of this ordering corresponds to the atomic execution of a_i.

If the execution of a_i^1 in the clock cycle c does not violate the peak power constraint of the design then a_i^1 can be executed in c. The other part a_i^2 can be executed in the clock cycle after c. However, for a_i^2 to execute in the clock cycle after c, two conditions need to be satisfied. First, none of the other actions executing in c should update the state elements accessed by a_i^2. Second, none of the actions executing with a_i^2 in the cycle after c should access the state elements updated by a_i^2. To satisfy the first condition, those actions are chosen for the factorization which are not dependent on the other actions executing in the clock cycle c. Second condition is satisfied by executing only those actions in the cycle after c which are not dependent on part a_i^2.

Similar to Algorithm 3, algorithms based on the above strategy also re-schedules the actions of the design based on the dependency relationships among various actions. Apart from the atomic actions considered for scheduling in Algorithm 3, these algorithms also consider parts of the factorized actions for scheduling in each clock cycle. Thus, additional dependency constraints associated with the parts of an action are also taken into account while re-scheduling for peak power minimization.

Example

Consider again the example shown in Fig. 5.4. Since action a_1 cannot be executed in the clock cycle c, the strategy based on the factorization of the actions of the design will divide the action a_1 into parts a_1^1 and a_1^2. Action a_1 is factorized such that the atomic execution of a_1 corresponds to the behavior corresponding to the execution of a_1^1 followed by a_1^2.

The peak power constraint does not allow three or more actions to execute in the same clock cycle. The schedule obtained by re-scheduling the factorized parts of a_1 under these peak power constraints is shown in Fig. 5.6. There, the part a_1^1 is scheduled to execute in clock cycle c and the part a_1^2 in postponed to execute in the clock cycle after c. Action a_4 is scheduled to execute concurrently with part a_1^2 because it is dependent only on the part a_1^1 which is scheduled to execute in clock cycle c. Since action a_1 is factorized into two parts which execute in different clock cycles, therefore the peak power constraint of the design in not violated. It should be noticed that using this strategy, the latency of the design remains same as the base-line schedule shown in Fig. 5.4.

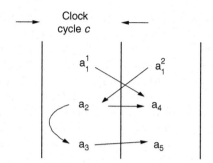

Fig. 5.6 Schedule using the factorization of actions under peak power constraints

5.2.2.2 Dynamic Power Reduction

Factorization of the actions of a design can also be used to reduce the switching activity of the design, thus reducing its dynamic power consumption. An action that involves a state element in the same operation more than once can be subdivided into parts such that these similar operations (using the same state element) exist in different parts. Executing these parts in consecutive clock cycles and assigning same functional units to such operations may result in the reduction of the dynamic power consumption of the design. As mentioned earlier, an action should be factorized such that there exists at least one sequential ordering of the execution of its parts whose behavior corresponds to the atomic execution of that action.

Let us assume that two such parts of an action a_i be a_i^1 and a_i^2. Also assume that in the base-line schedule, a_i is supposed to execute in a clock cycle c. Using this strategy, only a_i^1 is executed in cycle c and a_i^2 is postponed to the clock cycle after c. However, two constraints need to be satisfied to schedule the execution of a_i^2 in the cycle after c. First, none of the other actions executing in c should update the state elements accessed by a_i^2. Second, none of the actions executing with a_i^2 in the cycle after c should access the state elements updated by a_i^2. Any actions violating these conditions can be postponed to the future clock cycles by defining a conflict among appropriate actions of the designs.

5.2.3 Functional Equivalence

The overall execution of a design can usually be described as the execution of a collection of one or more transactions. Depending on the functionality of the design, these transactions can be simple or complicated. An example of a simple transaction can be the reading of a value from a FIFO which can be completed in one clock cycle. On the other hand, some designs may involve much more complicated transactions like processing of a request. Such transactions will take multiple cycles for the complete execution. Synchronization points of a design are usually expressed in terms of the completion of such transactions describing the design. In other words, a particular synchronization point of a design corresponds to the completion of one or more of its transactions. Given the synchronization points of a design, the schedule of the design obtained by using the low-power strategies should be equivalent to the base-line schedule (where the maximal set of actions are executed in each clock cycle without considering any power savings) at these points. For Algorithm 1, identification of such synchronization points is straightforward since the algorithm does not consider any new actions for execution until executing all the enabled actions of a clock cycle. Thus, schedule obtained after using Algorithm 1 can be easily proved to be functionally equivalent to the base-line schedule. However, for Algorithm 3 such equivalence is not straightforward since it considers actions over multiple clock cycles. Below, we sketch a proof for functional equivalence between the new schedule obtained on applying Algorithm 3 and the base-line schedule.

Proposition: Given a CAOS-based design and its base-line schedule γ, all the strategies based on the re-scheduling and the factorization of the actions proposed in Section 5.2 result in a new schedule θ that is functionally equivalent to γ, assuming that appropriate input mapping is done to feed the two schedules corresponding inputs at the right times.

Proof Sketch: Let the base-line schedule of a design be denoted as $\gamma = \{A_1, A_2, \ldots, A_n, \ldots\}$. Let $A_c = \{a_{c1}, a_{c2}, \ldots, a_{cm}\}$ denote the set of actions executing in the clock cycle c of schedule γ. Suppose that the concurrent execution of these actions corresponds to the given sequential ordering of the actions in A_c. Let us consider a new schedule in which the cycle c of the base-line schedule is replaced by the set $C = \{c_1, c_2, \ldots, c_m\}$ of m consecutive clock cycles each executing only one action. The actions are assigned to these clock cycles in the order given in A_c. We call this new schedule as the "sequential schedule" and it is denoted as $\omega = \{A_1', A_2', \ldots, A_r', \ldots\}$, where $A_i' = a_i, a_i \in A$. The amount of execution performed in the single clock cycle c of the base-line schedule is same as performed in the m clock cycles (from c_1 to c_m) of set C in the sequential schedule. Therefore, the sequential schedule is functionally equivalent to the base-line schedule at various synchronization points.

A_1 denotes the corresponding sequential ordering of the actions executing in the first cycle of the base-line schedule. Let A_1 contain d number of actions. Consider a new schedule θ generated using one of the low-power strategies proposed in Section 5.2. Starting from the first element of A_1 and following the ordering in A_1, the low-power strategy chooses a set $A_1' \subseteq A_1$ of b actions to execute in the first clock cycle, where $(b \leq d)$. Any remaining set of actions $A_1'' \subset A_1$ such that $A_1'' \cap A_1 = \phi$ are postponed to execute in the future clock cycles. In the second clock cycle, let A_s be the ordered set of actions whose guards evaluated to *True* and that do not conflict with each other.

If b equals d, then after the first clock cycle, computation done in schedule θ will be the same as the computation done in the first d clock cycles of the sequential schedule ω.

If $b < d$, then in the second clock cycle, the low-power strategy first considers (in the given order) all the actions in $A_1^{''}$ for execution. It also considers all those actions of A_s that do not violate the dependency relationships with the actions in $A_1^{''}$ or conflicts with those actions of $A_1^{''}$ which are chosen for execution in the second clock cycle. After the second clock cycle, the computation done in schedule θ will at least be the same as the computation done in the first c clock cycles of the sequential schedule ω, where $c \in \{b+1, \ldots d\}$. Some extra computation (depending on the number of actions of A_s executed in the second clock cycle) which does not conflict with the other computation might also occur. The actions of $A_d^{''}$ and A_s that are not allowed to execute in the second clock cycle are considered for execution in the third clock cycle in the appropriate order. The execution of the actions continues so on and thus, eventually after some clock cycles, say p, the computation done in schedule θ will at least be the same as the computation done in the first d clock cycles of the sequential schedule ω. Therefore, the end of the clock cycle p will denote the point where the state in the low-power schedule is equivalent to the state after d clock cycles of the sequential schedule have executed. Similarly, for the first synchronization point of the sequential schedule, a clock cycle will exist in the low-power schedule where the state at the end of that cycle is equivalent to the state at that point in the sequential schedule.

Now, lets assume that there also exists a clock cycle, say q, in the low-power schedule at the end of which the state of the design is equivalent to the state at the kth synchronization point of the sequential schedule. Assume that the kth synchronization point is located at the end of the tth clock cycle of the sequential schedule.

In the clock cycles after q, the low-power schedule will execute the actions which correspond to some clock cycles after t in the sequential schedule. Based on the similar analysis as above, it can be claimed that eventually after some more clock cycles in the low-power schedule the state of the design will be equivalent to the state at the $(k+1)$th synchronization point of the sequential schedule.

Thus, using mathematical induction, it can be claimed that for each synchronization point of the sequential schedule a clock cycle will exist in the low-power schedule such that the state at the end of that cycle is equivalent to the state at that synchronization point in the sequential schedule. The execution of the actions based on the orderings ensures that these clock cycles of the low-power schedule follow the sequence of those synchronization points. Therefore, the low-power schedule is always functionally equivalent to the sequential schedule at these synchronization points. Since the sequential schedule is functionally equivalent to the base-line schedule at the same set of synchronization points, we can say that the low-power schedule is functionally equivalent to the base-line schedule having maximal concurrency.

5.2.4 Example Applications

5.2.4.1 Peak Power Reduction

Consider again the LPM (longest prefix match) module design of Fig. 4.1. As mentioned earlier, based on the number of operations of in each action, *Action 1*, *Action 2*, and *Action 3* are assigned weights of 4, 3, and 4, respectively.

In the base-line schedule, *Action 1* and *Action 2* can be executed in the same clock cycle under no peak power constraints. Let us consider one such clock cycle c as shown in Fig. 5.7. Assume that the maximum allowable peak power for the LPM module is 4 units. The concurrent execution of *Action 1* and *Action 2* in the same clock cycle results in 7 units of power consumption in cycle c and thus, violates the peak power constraint of the design.

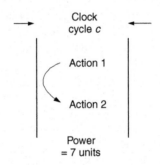

Fig. 5.7 Schedule under no peak power constraints (LPM design)

Using Algorithm 3 under the peak power constraint of 4 units will generate the schedule shown in Fig. 5.8. The algorithm re-schedules the execution of the design based on their dependency relationships such that the peak power constraint of the design is satisfied. Since *Action 2* is dependent on *Action 1*, therefore *Action 1* is scheduled in the next clock cycle to meet the peak power constraint of the design.

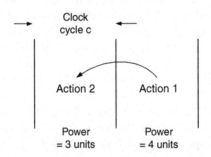

Fig. 5.8 Schedule by applying Algorithm 3 when $P_{peak} = 4$ (LPM design)

Now, let us assume that the maximum allowable peak power of the design is reduced to 3 units. Under this constraint, Algorithm 3 fails to produce a feasible schedule of the design since *Action 1* consumes 4 units of power which is more than the maximum allowable peak power. Applying the algorithm based on the factorization of the actions to generate the schedule under the peak power constraint of 3 units generates the schedule shown in Fig. 5.9. The algorithm factorizes *Action 1* into two parts *Action 1.1* and *Action 1.2* each of which consumes 2 units of power. The algorithm then schedules these parts to execute in different clock cycles. This shows that, under a tight peak power constraint, algorithm based on the factorization of actions may be able to produce a feasible schedule when the other algorithm, which does not factorizes the actions is unable to do so.

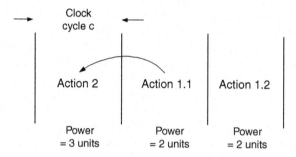

Fig. 5.9 Schedule using the factorization of actions when $P_{peak} = 3$ (LPM design)

5.2.4.2 Dynamic Power Reduction

For the dynamic power minimization problem, the strategy based on the factorization of the actions targets the reduction of the switching activity in a design by allowing the sharing of functional resources not only among different actions of a design but also among the parts of these actions. This may result in some additional power savings as compared to the strategy which does not factorizes the actions of a design. Thus, the amount of power reduction achieved from strategies based on factorization may be more.

5.2.4.3 Effects on Latency

1. *When considering peak power reduction*
   When the peak power constraint of a design is specified, the low-power strategy re-schedules the execution of the design based on that constraint. The peak power constraint does not allow maximal set of actions to execute in a clock cycle, thus increasing the latency of the design while meeting the peak power constraint. The strategy based on the factorization of the actions looks for the opportunity to execute the parts of an action in a clock cycle c in which the action itself cannot be executed atomically. In the clock cycle after c, any actions dependent on the parts executed in c can be considered for execution. When factorization of the actions is not used, all such actions cannot be considered for execution in the

clock cycle after c. Thus, in most cases, the factorization of the actions for peak power reduction improves the latency of the design since more computation takes place in each clock cycle while meeting the peak power constraints of the design.

When the latency constraint of a design is specified, the average peak power constraint of the design can be estimated using Eqs. (4.3) and (4.4) of Chapter 4. This estimated peak power constraint can be used to re-schedule the execution of the design in order to meet its latency constraint. As mentioned earlier, the strategy based on the factorization of the actions performs more computation in each clock cycle and hence provides better achievement of the latency constraint.

2. *When considering dynamic power reduction*
   The re-scheduling of the actions for dynamic power reduction reduces the switching activity of the design by scheduling appropriate actions in the different clock cycles. If such actions are executed in the same clock cycle in the baseline schedule, this re-scheduling of the actions in the different clock cycles will increase the latency of the design. In addition to re-scheduling of the actions of a design, the strategy based on the factorization of the actions also re-schedules the parts of an action to different clock cycle in order to achieve further reduction in dynamic power. Thus, in most cases, this factorization of the actions used for dynamic power savings will further increase the latency of the design.

Chapter 6
Complexity Analysis of Scheduling in CAOS-Based Synthesis

Scheduling, allocation and binding are three important phases of a CDFG-based synthesis process. These phases are interdependent and can be performed in different orders depending on the design flow. In some cases, two or more phases can also be performed simultaneously. Such simultaneous execution of these phases during a synthesis process will result in a solution which is globally optimal. However, the problem of performing the three phases simultaneously is *NP*-hard. For this reason, most CDFG-based synthesis flows perform these phases separately in order to increase the possibility of finding optimal polynomial time algorithms for each phase.

Scheduling affects allocation and binding phases and hence efficient scheduling of the operations of a design is very important in producing designs that are efficient in terms of area, latency, and power. In general, the resource-constrained and time-constrained versions of the scheduling problem are known to be *NP*-complete [17, 116]. The best known algorithms for solving these problems have exponential time complexity, and hence heuristics are often used to solve such problems.

In this chapter, we perform a detailed analysis of the complexity of scheduling problems associated with the CAOS-based synthesis process. We also focus on the peak power requirements of a design and propose several other heuristics (in addition to the heuristics presented in Chapter 5) for efficient scheduling of the actions of a CAOS-based design such that its peak power constraints are satisfied. We discuss the complexity and approximability of a variety of problems involving peak power and schedule length encountered in the CAOS-based synthesis. Exploiting the relationships between these problems and classical optimization problems such as bin packing and multiprocessor scheduling, we show how one can develop efficient approximation algorithms with provable performance guarantees.

We first present the complexity analysis of the scheduling problem associated with the CAOS-based synthesis without any peak power constraints in Section 6.2. This paves the way to the analysis under peak power constraint in the next section. In Section 6.3, various versions of the scheduling problem under peak power constraints are discussed and efficient heuristics are presented in each case. We also discuss how such approximations can be applied to meet the peak power goal in designs generated using *Bluespec Compiler (BSC)*, which is a commercial high-level synthesis tool based on the CAOS (by *Bluespec Inc.*).

G. Singh, S.K. Shukla, *Low Power Hardware Synthesis from Concurrent Action-Oriented Specifications*, DOI 10.1007/978-1-4419-6481-6_6,
© Springer Science+Business Media, LLC 2010

6.1 Related Background

6.1.1 Confluent Set of Actions

A CAOS-based design can be composed of conflict-free actions. Two actions are said to be conflict-free if none of the actions updates the state elements accessed (including the state elements accessed in the guards) by the other action [62]. A set of actions where all the pairs of actions are conflict-free is said to be *confluent* [11]. The important thing to note about a confluent set of actions is that all the possible orders of the execution of such actions will result in the same final state of the design.

6.1.2 Peak Power Constraint

As mentioned earlier, during the execution of a CAOS-based design, multiple non-conflicting actions can be selected for execution from all the actions enabled (guards evaluated to *True*) in a particular time slot (clock cycle). Let A be such a set of non-conflicting actions. Each action $a_i \in A$ is composed of a set of operations and the power expended during each operation of a_i contributes to power p_i needed to execute a_i.

Total power consumed during a particular time slot can be estimated as the summation of the power consumed by various actions executing in that time slot. Thus, executing large number of actions in a time slot may increase the peak power of the design beyond the acceptable limits, which is undesirable. A peak power constraint P on a design can be specified by imposing a limit on the total power that can be consumed in any time slot during the execution of the design. The scheduling of various actions of the design should then be done such that the peak power constraint of the design is not violated.

Note: For the rest of this chapter, we use P to denote the peak power constraint of a design.

6.2 Scheduling Problems Without a Peak Power Constraint

This section considers two scheduling problems related to CAOS-based synthesis in which there is no constraint on the amount of power that can be used in any time slot; that is, no peak power constraint exists. The first problem (considered in Section 6.2.1) is to construct a largest subset of non-conflicting actions from the set of actions enabled in a time slot. The second problem (considered in Section 6.2.2) concerns the construction of a minimum length schedule for all the actions.

6.2.1 Selecting a Largest Non-conflicting Subset of Actions

During the execution of a CAOS-generated design, two actions conflicting with each other cannot be executed in the same time slot. Given a set of actions enabled in a time slot and pairs of conflicting actions, this section considers the problem of finding a largest subset of pairwise non-conflicting actions. The idea is that all the actions in such a subset can be scheduled in the same time slot. We call it the *Maximum Non-conflicting Subset of actions (MNS)* problem and its formal definition is presented below.

Maximum Non-conflicting Subset of Actions (MNS)

Instance: A set $A = \{a_1, a_2, \ldots, a_n\}$ of actions; a collection C of pairs of actions, where $\{a_i, a_j\} \in C$ means that actions a_i and a_j conflict; that is, they cannot be scheduled in the same time slot; an integer $K \le n$.

Question: Is there subset $A' \subseteq A$ of such that $|A'| \ge K$ and no pair of actions in A' conflict?

In subsequent sections, we present complexity and approximation results for the MNS problem.

6.2.1.1 Complexity Results for the General Case

The following result points out that the MNS problem is, in general, computationally intractable. In particular, the result points out that the MNS problem corresponds to the well-studied MAXIMUM INDEPENDENT SET problem for undirected graphs.

Proposition 6.1 *The MNS problem is NP-complete.*

Proof The MNS problem is in *NP* since one can guess a subset A' of A and verify in polynomial time that $|A'| \ge K$ and that no pair of actions in A' conflict.

To show that MNS is *NP*-hard, we use reduction from the MAXIMUM INDEPENDENT SET (MIS) problem which is known to be *NP*-complete [54]. An instance of the MIS problem consists of an undirected graph $G(V, E)$ and an integer $J \le |V|$. The question is whether G has an independent set V' of size $\ge J$ (i.e., a subset V' of V such that $|V'| \ge J$ and there is no edge between any pair of nodes in V').

The reduction is straightforward. Given an instance I of the MIS problem, we construct an instance I' of the MNS problem as follows. The set $A = \{a_1, a_2, \ldots, a_n\}$ of actions is in one-to-one correspondence with the node set V, where $n = |V|$. For each edge $\{v_i, v_j\}$ of G, we construct the pair $\{a_i, a_j\}$ of conflicting actions. Finally, we set $K = J$. Obviously, the construction can be carried out in polynomial time. From the construction, it is easy to see that each independent set of G corresponds to a non-conflicting set of actions and vice versa. Therefore, G has an independent set of size J if and only if there is a subset A' of size $K = J$ such that the actions in A' are pairwise non-conflicting. ∎

The above reduction shows that there is a direct correspondence between the MNS and MIS problems. Thus, for any $\rho \ge 1$, a ρ-approximation algorithm for the MNS problem can also be used as a ρ-approximation algorithm for the MIS problem. It is known that for any $\varepsilon > 0$, there is no $O(n^{1-\varepsilon})$-approximation algorithm for the MIS problem, unless $P = NP$ [60, 121]. Thus, we have the following observation.

Observation 6.1 *For any $\varepsilon > 0$, there is no $O(n^{1-\varepsilon})$-approximation algorithm for the MNS problem, unless $P = NP$.*   ∎

6.2.1.2 Approximation Algorithms for a Special Case of MNS

As mentioned above, a polynomial time approximation algorithm with a good performance guarantee is unlikely to exist for general instances of the MNS problem. However, for special cases of the problem, one can devise heuristics with good performance guarantees by exploiting the close relationship between the MNS and MIS problems. We now present an illustrative example.

Consider instances of the MNS problem in which every action conflicts with at most Δ other actions, for some constant Δ. A simple heuristic which provides a performance guarantee of $\Delta + 1$ for this special case of the MNS problem is shown in Fig. 6.1. Step 1 of this heuristic transforms the special case of the MNS problem into a special case of the MIS problem where the underlying graph has a maximum node degree of Δ. It is a simple matter to verify that each independent set of the resulting graph corresponds to a subset of pairwise non-conflicting actions. Step 3 describes a greedy algorithm which provides a performance guarantee of $\Delta + 1$ for the restricted version of the MIS problem [13, 59]. Because of the direct correspondence between independent sets of G and non-conflicting subsets of actions, the heuristic also serves as a $(\Delta + 1)$-approximation algorithm for the special case of the MNS problem. This result is stated formally below.

Observation 6.2 *For instances of the MNS problem in which each action conflicts with at most Δ other actions for some constant Δ, the approximation algorithm in Fig. 6.1 provides a performance guarantee of $\Delta + 1$.*   ∎

---

1. From the given instance of the MNS problem, construct a graph $G(V,E)$ as follows: The node set V is in one-to-one correspondence with the set of actions A. For each conflicting pair $\{a_i, a_j\}$ of actions, add the edge $\{v_i, v_j\}$ to E.
2. Initialize A' to $\emptyset$. (**Note:** At the end, A' will contain a set of pairwise non-conflicting actions.)
3. **while** $V \neq \emptyset$ **do**

   (a) Find a node v of minimum degree in G.
   (b) Add the action corresponding to node v to A'.
   (c) Delete from G the node v and all nodes which are adjacent to v. Also delete the edges incident on those nodes.
   (d) Recompute the degrees of the remaining nodes.

4. Output A'.

Fig. 6.1 Steps of the heuristic for the special case of the MNS problem

For the class of graphs whose maximum node degree is bounded by a constant Δ, a heuristic which provides a better (asymptotic) performance guarantee of $O(\Delta \log\log\Delta / \log\Delta)$ is known for the MIS problem [13, 59]. However, that heuristic is harder to implement compared to the one shown in Fig. 6.1.

Efficient approximation algorithms with good performance guarantees are known for the MIS problem for other special classes of graphs such as planar graphs, near-planar graphs and unit disk graphs [12, 65, 66]. When the MNS instances correspond to instances of the MIS problem for such graphs, one can use the approximation algorithms for the latter problem to obtain good approximations for the MNS problem.

Application to Bluespec

Bluespec Compiler (BSC) allows a maximal set of non-conflicting actions to execute in each time slot which aids in reducing the latency of the design. Reference [80] presents an algorithm used in *BSC* which selects actions based on their priorities. The algorithm first orders the set A of all the actions in terms of their priorities. Let $A' = \{a_1, a_2, \ldots, a_n\}$ be the ordered set of actions such that a_i is more urgent (higher priority) than a_j iff $i < j$. The algorithm then computes a set S which contains all the actions that can be executed in a time slot as follows.

1. Let $S = \emptyset$.
2. If A' is empty, stop and return S.
3. Let a_k be the highest priority action in A'.
4. If a_k is enabled and no action a_j, $1 \le j < k$, conflicting with a_k is enabled, then $e = 1$, else $e = 0$.
5. If ($e == 1$) then add a_k to S.
6. Remove a_k from A'.
7. Go to step 2.

Thus, the above algorithm selects a set of non-conflicting actions based on their priorities. Between two conflicting actions, an action with higher priority is always chosen to execute whenever both the actions are enabled in the same time slot. If multiple actions conflict with each other, then the action with the highest priority is preferred for execution over all the other conflicting actions. Such a high-priority action is selected even if it conflicts with large number of other actions.

Note that step 3 of the heuristic shown in Fig. 6.1 selects the minimum degree node from the remaining set of nodes. Thus, during the selection of independent nodes, the heuristic gives low preference to a node connected to large number of other nodes. With regard to the MNS problem, this implies that if an action conflicts with large number of other actions then the heuristic gives a low preference to that action while selecting the set of non-conflicting actions. It gives priority to actions conflicting with minimum number of other actions, and thus attempts to select as many actions as possible. This may further reduce the latency of the design. On the other hand, in *BSC*, if the designer does not specify a priority between two conflicting actions, the compiler assigns an arbitrary priority to both of them. Thus, the algorithm used in *BSC* does not attempt to select a large set of non-conflicting actions, unlike the heuristic shown in Fig. 6.1.

If we change step 3 of the heuristic to select nodes based on their priorities, as done in *BSC*, then the result of the heuristic in Fig. 6.1 will be same as the algorithm used in *BSC*; that is, an enabled action with the highest priority will always be chosen to execute.

6.2.2 Constructing Minimum Length Schedules

The previous section considered the problem of choosing a non-conflicting set of maximum cardinality. Here, we consider the problem of partitioning a given set of actions into a minimum number of subsets so that no subset contains a pair of conflicting actions. When this condition is met, the actions in each subset can be scheduled in the same time slot. Thus, the partitioning problem models the problem of constructing a minimum length schedule for the set of actions. A formal definition of the scheduling problem is given below.

Minimum Length Schedule Construction (MLS)

Instance: A set $A = \{a_1, a_2, \ldots, a_n\}$ of actions; a collection C of pairs of actions, where $\{a_i, a_j\} \in C$ means that actions a_i and a_j conflict, that is, they cannot be scheduled in the same time slot; an integer $\ell \leq n$.

Question: Is there a partition of A into r subsets $A_1, A_2, \ldots, A_r$, for some $r \leq \ell$, such that for each i, $1 \leq i \leq r$, the actions in A_i are pairwise non-conflicting?

In what follows, we present complexity and approximation results for the MLS problem.

6.2.2.1 Complexity Results for the General Case

The following result points out that the MLS problem is, in general, computationally intractable. In particular, our result points out that the MLS problem corresponds to the *minimum coloring* problem for undirected graphs.

Proposition 6.2 *The MLS problem is NP-complete.*

Proof It is easy to see that the MLS problem is in *NP* since one can guess a partition of A into at most $r \leq \ell$ subsets and verify in polynomial time that no pair of actions in any subset conflict.

To show that MLS is *NP*-hard, we use reduction from the MINIMUM K-COLORING (K-COLORING) problem which is known to be *NP*-complete [54]. An instance of the K-COLORING problem consists of an undirected graph $G(V, E)$ and an integer $K \leq |V|$. The question is whether the nodes of G can be colored using at most K colors so that for each edge $\{v_i, v_j\} \in E$, the colors assigned to v_i and v_j are different. We note that in any valid coloring of G, each color class (i.e., the set of nodes assigned the same color) forms an independent set.

The reduction is straightforward. Given an instance I of the K-COLORING problem, we construct an instance I' of the MLS problem as follows. The set $A = \{a_1, a_2, \ldots, a_n\}$ of actions is in one-to-one correspondence with the node set V, where $n = |V|$. For each edge $\{v_i, v_j\}$ of G, we construct the pair $\{a_i, a_j\}$ of conflicting actions. Finally, we set $\ell = K$. Obviously, the construction can be carried out in polynomial time. From the construction, it is easy to see that each valid coloring set of G corresponds to a partition of the set A into non-conflicting subsets and vice versa. (Each color class corresponds to a subset of actions that can be scheduled in the same time slot and vice versa.) Therefore, G has a valid coloring with $r \leq K$ colors if and only if there is a partition of A into r subsets such that the actions in each subset are pairwise non-conflicting.   ∎

Let us denote the problem of coloring a graph with a minimum number of colors by MINCOLOR. The above reduction shows that there is a direct correspondence between the MLS and MINCOLOR problems. Thus, for any $\rho \geq 1$, a ρ-approximation algorithm for the MLS problem can also be used as a ρ-approximation algorithm for the MINCOLOR problem. It is known that for any $\varepsilon > 0$, there is no $O(n^{1-\varepsilon})$-approximation algorithm for the MINCOLOR problem, unless $P = NP$ [51, 121]. Thus, we have the following observation.

Observation 6.3 *For any $\varepsilon > 0$, there is no $O(n^{1-\varepsilon})$-approximation algorithm for the MLS problem, unless $P = NP$.*   ∎

6.2.2.2   Observations Concerning Special Cases of MLS

As mentioned above, the MLS problem is, in general, hard to approximate. To identify some simpler versions of the MLS problem, we exploit the relationship between the MLS and MINCOLOR problems. In particular, the MLS problem can be reduced to the MINCOLOR problem by constructing a graph G in which each node corresponds to an action and each edge corresponds to a pair of conflicting actions. It is easy to see that any valid coloring of G with r colors corresponds to a schedule of length r. This reduction of MLS to the MINCOLOR problem is useful for several reasons. First, it points out that in practice, one can use known heuristics for graph coloring in constructing schedules of near-minimum length. Although the coloring problem is hard to approximate in the worst-case [51], heuristics that work well in practice are known (see for example [21, 61]). In addition, the reduction to the MINCOLOR problem also points out that the following two special cases of the MLS problem can be solved efficiently.

1. Consider instances of the MLS problem in which the upper bound on the length of the schedule is two. This special case of the MLS problem corresponds to the problem of determining whether a graph is 2-colorable. Efficient algorithms are known for this problem [41].

2. Consider instances of the MLS problem in which each action conflicts with at most Δ other actions, for some integer Δ. For such instances, a schedule of length at most $\Delta + 1$ can be constructed in polynomial time. To see this, we note that the graph corresponding to such instances of the MLS problem has a maximum node degree of Δ. By a well-known result in graph theory, called Brooks's Theorem, any graph with a maximum node degree of Δ can be colored efficiently using at most $\Delta + 1$ colors [118]. Such a coloring gives rise to a schedule of length at most $\Delta + 1$.

Application to Bluespec

In *Bluespec*, multiple conflicting actions can get enabled in each time slot. But only a non-conflicting subset of such actions can be allowed to execute in a given time slot. Thus, the heuristics for MLS problem can be used for partitioning such a set of actions into multiple subsets of non-conflicting actions. Each of these subsets can then be scheduled in a separate time slot. The two special cases of the MLS problem discussed above may also occur frequently in *Bluespec* designs. Corresponding algorithms suggested above can then be applied to efficiently solve such cases.

Note that executing actions of one subset may disable (guards evaluate to *False*) the actions of other subsets. This may happen if the state elements updated by the actions belonging to a subset (executing in the present time slot) are accessed by the guards of the actions belonging to the other subsets (which are scheduled to execute in future time slots). However, if the guards of these actions of other subsets do not access the state elements updated by the subset of actions executing in the present time slot, then such partitioning of actions can be used in *Bluespec* to schedule each non-conflicting subset in a different time slot. A set of actions is called a *persistent set* if executing a subset of these actions does not disable the guards of the remaining actions of the set. Thus, in general, such partitioning of actions can be used to schedule a persistent set of actions.

6.3 Scheduling Problems Involving a Power Constraint

In this section, we study scheduling problems taking into consideration the amount of power consumed by various actions. In particular, we assume that a constraint on peak power, that is, the amount of power that can be consumed in any time slot, is specified. The goal is to construct schedules satisfying this constraint. We first consider (Section 6.3.1) the problem of finding a largest subset of actions from a set of non-conflicting actions that can be scheduled in a given time slot under the peak power constraint. We show that this problem can be solved efficiently. Next, a generalization of this problem, where there is a utility value associated with each action and the goal is to choose a subset of actions that maximize the total utility while satisfying the peak power constraint is shown to be *NP*-complete (Section 6.3.2). Finally (Section 6.3.3), we consider several versions of the problem of minimizing the schedule length subject to the peak power constraint. We present both complexity and approximation results.

6.3.1 Packing Actions in a Time Slot Under Peak Power Constraint

Due to the peak power constraint, it might not be possible to execute all the actions belonging to a set of non-conflicting actions that are enabled in a particular time slot. This section considers the problem of packing a maximum number of actions into a time slot without violating the constraint on peak power. We present a simple algorithm with a running time of $O(n \log n)$ for the problem. We begin with a formal statement of the problem.

Maximum Number of Actions in a Time Slot Subject to Peak Power Constraint (MNA-PP)

Instance: A set $A = \{a_1, a_2, \ldots, a_n\}$ of non-conflicting actions; for each action a_i, the power p_i needed to execute that action; a positive number P representing the peak power, that is, the maximum power that can be used in any time slot.

Requirement: Find a subset $A' \subseteq A$ such that the total power needed to execute all the actions in A' is at most P and $|A'|$ is a maximum over all subsets of A that satisfy the constraint on peak power.

An efficient algorithm for MNA-PP is shown in Fig. 6.2. The following lemma shows the correctness of the algorithm.

---

1. Sort the actions in A into non-decreasing order by the amount of power needed for each action. Without loss of generality, let $\langle a_1, a_2, \ldots, a_n \rangle$ denote the resulting sorted order.
2. **Comment:** Keep adding actions in the above order as long as the total power is constraint is satisfied.

    (a) Initialization: Let $A' = \emptyset$, $i = 1$ and $R = P$. (**Note:** R denotes the remaining amount of power.)
    (b) **while** $(i \le n)$ **and** $p_i \le R$ **do**
        (i) Add a_i to A'.
        (ii) $R = R - p_i$.
        (iii) $i = i + 1$.

3. Output A'.

Fig. 6.2 Steps of the algorithm for the MNA-PP problem

Lemma 6.1 *The algorithm in Fig. 6.2 computes an optimal solution to the MNA-PP problem.*

Proof Let $\langle a_1, a_2, \ldots, a_n \rangle$ denote the list of actions in nondecreasing order of power values. We assume that P is sufficient to execute a_1; otherwise, none of the actions can be executed in any time slot. Let A' be the set of actions produced by the algorithm in Fig. 6.2, and let $|A'| = k$. Thus, $A' = \{a_1, a_2, \ldots, a_k\}$, and the remaining power is not enough to add action a_{k+1} to A'.

Suppose A' is not an optimal solution. Thus, there is another solution A^* such that $|A^*| = r \geq k+1$. Let $\langle a_{i_1}, a_{i_2}, \ldots, a_{i_r} \rangle$ denote the actions in A^* arranged in the sorted order chosen by the algorithm. It is easy to see that, for $1 \leq j \leq k$, the power needed to execute action a_j is no more than that of a_{i_j}. Thus, the remaining power after adding actions $a_{i_1}, a_{i_2}, \ldots, a_{i_k}$ is no more than that after adding the actions $a_1, a_2, \ldots, a_k$. Further, the power needed for action $a_{i_{k+1}}$ is at least that needed for action a_{k+1}. Thus, the optimal solution A^* cannot accommodate the action $a_{i_{k+1}}$ without violating the peak power constraint. This is a contradiction and the lemma follows. ∎

Theorem 6.1 *The* MNA-PP *problem can be solved in* $O(n \log n)$ *time, where n is the number of actions.*

Proof The correctness of the algorithm in Fig. 6.2 follows from Lemma 6.1. To estimate the running time of the algorithm, we note that the sorting step uses $O(n \log n)$ time and the other steps use $O(n)$ time. So, the overall running time is $O(n \log n)$. ∎

Application to Bluespec

As already mentioned, in *Bluespec*, a maximal subset of non-conflicting actions are executed in each time slot. If a large number of actions are executed in the same time slot, then it can lead to the violation of peak power constraint of the design. In that case, only a subset of such actions should be allowed to execute in a given time slot to meet the peak power constraint.

Thus, in *Bluespec*, the problem of selecting a largest subset of actions such that the peak power constraint of the design is satisfied can be directly mapped to the MNA-PP problem described above. Algorithm in Fig. 6.2 can then be used to solve the peak power problem optimally.

In the real hardware, solving MNA-PP problem in each time slot during the execution of a design might consume some extra power. To avoid such power overhead, a small associative memory element can be used to decide what subset of the enabled actions is allowed to execute in a particular time slot. The memory element can be used to store various combinations of the actions which will result in the peak power violation when executed in a time slot. For each such combination, the memory can return a corresponding subset of actions which satisfies the peak power constraint. Thus, in a particular time slot, if the enabled set of non-conflicting actions results in the peak power violation, then the memory can be used to return a suitable subset of actions which can be executed to meet the peak power requirements.

Note that if the number of combinations is not large, then instead of memory element, a lookup-table can also be used to store the combinations. One such lookup table-based synthesis approach is presented in Chapter 8 along with the corresponding experimental results.

6.3.2 *Maximizing Utility Subject to a Power Constraint*

We now consider a generalization of the MNA-PP problem considered in the previous subsection. Suppose for each action $a_i \in A$, there is a *utility* value u_i, in addition to the power value p_i. We can define the utility of a subset A' of actions to be the sum of the utilities of the actions in A'. It is of interest to consider the problem of selecting a subset of maximum utility, subject to the constraint on peak power. A formal statement of the decision version of this problem is as follows.

Maximizing Utility Subject to Peak Power Constraint (MU-PP)

Question: A set $A = \{a_1, a_2, \ldots, a_n\}$ of non-conflicting actions; for each action a_i, the power p_i consumed during the execution of that action and the utility u_i of that action; a positive number P representing the peak power; a positive number Γ representing the required utility.

Question: Is there a subset $A' \subseteq A$ such that the total power needed to execute all the actions in A' is at most P and the utility of A' is at least Γ?

Note that the MNA-PP problem is a special case of the MU-PP problem, with the utility value each action being 1. However, while the MNA-PP problem is efficiently solvable, the MU-PP problem is *NP*-complete, as shown below.

Proposition 6.3 *The* MU-PP *problem is NP-complete.*

Proof The MU-PP problem is in *NP* since one can guess a subset A' of A and verify in polynomial time that the utility of $|A'|$ is at least Γ and that the total power needed to execute the actions in A' is at most P.

To show that MU-PP is *NP*-hard, we use reduction from the KNAPSACK problem which is known to be *NP*-complete [54]. An instance of the KNAPSACK problem consists of a set S of items, an integer weight w_i and an integer profit value u_i for each item $s_i \in S$, and two positive integers B and Π. The question is whether there is a subset S' of S such that the total weight of all the items in S' is at most B and the profit of all the items in S' is at least Γ.

The reduction is straightforward. Given an instance I of the KNAPSACK problem, we construct an instance I' of the MU-PP problem as follows. The set $A = \{a_1, a_2, \ldots, a_n\}$ of actions is in one-to-one correspondence with the set S, where $n = |S|$. For each action a_i, the power value p_i and the utility u_i are set, respectively, to the weight w_i and the profit u_i of the corresponding item $s_i \in S$, $1 \le i \le n$. Finally, we set the power bound $P = B$ and the utility bound $\Gamma = \Pi$. Obviously, the construction can be carried out in polynomial time. From the construction, it is easy to see that any solution to the KNAPSACK problem with a profit of α corresponds to a set of actions whose utility is α and vice versa. Therefore, there is a solution to the KNAPSACK instance I if and only if there is a solution to the MU-PP instance I'.   ∎

Approximation Algorithms for MU-PP

It is easy to see that the optimization version of the MU-PP problem can be transformed into the KNAPSACK problem. Each action a_i is represented by an item s_i; the utility u_i and the power value p_i are, respectively, the weight and the profit values for the item. The peak power value P represents the knapsack capacity. Because of this transformation, known algorithms for the KNAPSACK problem can be directly used to solve the MU-PP problem. For example, a pseudo-polynomial algorithm for the MU-PP problem follows from the corresponding algorithm for the KNAPSACK problem [54]. Further, any approximation algorithm for the KNAPSACK problem can be used as an approximation algorithm with the same performance guarantee for the optimization version of MU-PP. For example, a known 2-approximation algorithm for KNAPSACK [54] implies a similar approximation for the optimization version of the MU-PP problem. Also, when the weights and profits are integers, there is a polynomial time approximation scheme (PTAS) for the KNAPSACK problem [54]. Therefore, when the power and utility values for each action are integers, one can obtain a PTAS for the optimization version of the MU-PP problem.

Application to Bluespec

The utility value of a given action can be used to represent the usefulness of the execution of that action. In *Bluespec*, an action $a_i \in A$ is said to be dependent on another action $a_j \in A$ if a_j updates a state element accessed by a_i [107]. Thus, in *Bluespec*, one measure of the utility value of an action a_j is the number of actions having a dependency on it; that is, the number of actions accessing the state elements updated by a_j. Based on such a measure, an action having a large number of other actions dependent on it will be assigned a high utility value.

Hence, when the actions of a *Bluespec* design are assigned different utility values, approximation algorithms for the MU-PP problem described above can be used to solve the corresponding *Bluespec* peak power problem – selecting a subset $A' \subseteq A$ such that the total power needed to execute all the actions in A' is at most P and the utility of A' is at least Γ.

6.3.3 Combination of Makespan and Power Constraint

Peak power constraint of a design may not allow all the actions in a set of non-conflicting actions to execute in a single time slot. In this subsection, our focus is on scheduling such a set of actions over a small number of time slots while keeping the peak power value as small as possible. The number of slots used by a schedule is called the *makespan*. Two optimization problems can be studied in this context. First, the problem of minimizing makespan subject to a peak power constraint can be formulated as follows.

Minimizing Makespan Subject to Peak Power Constraint (MM-PP)

Instance: A set $A = \{a_1, a_2, \ldots, a_n\}$ of non-conflicting actions; for each action a_i, the power p_i needed to execute that action; a positive number P representing the peak power, that is, the maximum power that can be used in any time slot.

Requirement: Find a schedule of minimum length for the actions in A such that the total power needed to execute the actions in each time slot is at most P.

The dual problem of minimizing peak power subject to a constraint on the schedule length can be formulated as follows.

Minimizing Peak Power Subject to a Makespan Constraint (MPP-M)

Instance: A set $A = \{a_1, a_2, \ldots, a_n\}$ of non-conflicting actions; for each action a_i, the power p_i needed to execute that action; a positive integer L representing the makespan.

Requirement: Find a schedule of length at most L for the actions in A such that the maximum total power used in any time slot is a minimum over all schedules of length at most L.

We note that the decision versions of the two problems MM-PP and MPP-M are identical. A formal statement of the decision version is as follows.

Minimizing Makespan and Peak Power – Decision Version (MPP-DECISION)

Instance: A set $A = \{a_1, a_2, \ldots, a_n\}$ of non-conflicting actions; for each action a_i, the power p_i needed to execute that action; a positive number P representing the peak power that can be used in any time slot; a positive integer L representing the makespan.

Question: Is there a schedule of length at most L for the actions in A such that the total power used in any time slot is at most P?

We now show that MPP-DECISION is *NP*-complete, even when the makespan is fixed at 2.

Proposition 6.4 *Problem* MPP-DECISION *is NP-complete.*

Proof The MPP-DECISION problem is in *NP* since one can guess a schedule for the actions in A and verify in polynomial time that the schedule length and the peak power constraints are satisfied.

To show that MPP-DECISION is *NP*-hard, we use reduction from the PARTITION problem which is known to be *NP*-complete [54]. An instance of the PARTITION problem consists of a set $S = \{s_1, s_2, \ldots, s_n\}$ of n integers. The question is whether

S can be partitioned into two subsets S_1 and S_2 such that the sum of the elements in S_1 is equal to the sum of the elements in S_2. (We may assume without loss of generality that $\sum_{i=1}^{n} s_i$ is even; otherwise, there is no solution to the PARTITION problem.)

Given an instance I of the PARTITION problem, we construct an instance I' of the MPP-DECISION problem as follows. The set $A = \{a_1, a_2, \ldots, a_n\}$ of actions is in one-to-one correspondence with the set S of items. For each action a_i, the power value p_i is set equal to s_i, $1 \leq i \leq n$. Finally, we set the bound P on peak power to $(\sum_{i=1}^{n} s_i)/2$ and the schedule length L to 2. This completes the construction. Obviously, the construction can be carried out in polynomial time. We now show that the PARTITION instance I has a solution if and only if the MPP-DECISION instance I' has a solution.

Part 1: Suppose the PARTITION instance I has a solution given by subsets S_1 and S_2. Consider the schedule of length 2 which includes all the actions corresponding to the numbers in S_1 in the first time slot and the remaining actions in the second time slot. Since the sum of the integers in S_1 and S_2 are both equal to $P = (\sum_{i=1}^{n} s_i)/2$, the total power used in each time slot is equal to P. Thus, we have a schedule of length two satisfying the peak power constraint; in other words, there is a solution for the MPP-DECISION instance I'.

Part 2: Suppose the MPP-DECISION instance I' has a solution, that is, a schedule of length at most 2 such that in each time slot, the power used is at most P. We first note that the schedule cannot be of length 1; if so, the peak power used would be equal to $\sum_{i=1}^{n} s_i$, which is greater than P. Thus, the schedule is of length 2. For $i = 1, 2$, let S_i be the set of integers corresponding to the actions scheduled in time slot i. We claim that the sum of the integers in S_1 is equal to that of S_2. To see this, note that the sum of the integers in S_1 cannot exceed $P = (\sum_{i=1}^{n} s_i)/2$, since that will violate the peak power constraint in the first time step. Likewise, if the sum of the integers in S_1 is less than P, then the sum of the integers in S_2 would exceed P; that is, the peak power constraint would be violated in the second time step. Thus, sum of the integers in S_1 must be equal to that of S_2. In other words, we have a solution to the PARTITION instance I. This completes the proof of Proposition 6.4. ∎

The above proof shows that determining whether there is a schedule of length 2 satisfying the peak power constraint is *NP*-complete. In contrast, we note that determining whether there is a schedule of length 1 satisfying the peak power constraint is trivial; we need to only check whether the total power for all the actions in A is at most the peak power value.

While the above proof shows that the MPP-DECISION problem is *NP*-complete even for fixed values of schedule length, it leaves open the possibility of a pseudo-polynomial algorithm for the problem. We now present a different reduction to show that MPP-DECISION is *strongly NP*-complete, when the schedule length is part of the problem instance. Thus, in general, there is no pseudo-polynomial algorithm for the MPP-DECISION problem, unless $P = NP$.

Proposition 6.5 *Problem* MPP-DECISION *is strongly NP-complete when the schedule length is part of the problem instance.*

Proof We use a reduction from the 3-PARTITION problem, which is known to be strongly *NP*-complete [54]. An instance I of this problem consists of a positive integer B, a positive integer m, a set $S = \{s_1, s_2, \ldots, s_{3m}\}$ of $3m$ positive integers such that $B/4 < s_i < B/2$, $1 \le i \le m$, and $\sum_{i=1}^{3m} s_i = mB$. The question is whether the set S can be partitioned into m subsets such that the sum of the values in each subset is exactly B. (The constraint on the value of each $s_i \in S$ ensures that when there is such a partition, each subset in the partition has exactly three elements.)

Given an instance I of 3-PARTITION, an instance I' of MPP-DECISION can be constructed as follows. The set $A = \{a_1, a_2, \ldots, a_{3m}\}$ of actions is in one-to-one correspondence with the set S of items. For each action a_i, the power value p_i is set equal to s_i, $1 \le i \le n$. We set the bound P on peak power to B and the schedule length L to m. This completes the construction. Obviously, the construction can be carried out in polynomial time. The proof that the 3-PARTITION instance I has a solution if and only if the MPP-DECISION instance I' has a solution is similar to that presented in the proof of Proposition 6.4. ∎

6.3.4 Approximation Algorithms for MM-PP

Recall that in the MM-PP problem, we are required to minimize the schedule length subject to a constraint on the peak power value. Since this problem is *NP*-hard, it is of interest to investigate approximation algorithms with provable performance guarantees. One can obtain such approximation algorithms by reducing the problem to the well-known BIN PACKING problem and using known approximation algorithms for the latter problem.

In the BIN PACKING problem, we are given a collection C of n items, where item i has a size $x_i \in (0, 1]$. The goal is to pack these items into a *minimum* number of bins, each of unit capacity. This minimization problem is known to be *NP*-hard [54]. However, several approximation algorithms with good performance guarantees are known for this problem [39]. For example, the problem admits a PTAS. However, the algorithm is somewhat complicated and its running time is exponential in $1/\varepsilon$, where ε is the fixed accuracy parameter. A much simpler algorithm, called *First Fit Decreasing* (FFD), provides a performance guarantee of $11/9$ [39]. The idea is first to sort the items in non-increasing order of their sizes and then assign each item to the first bin in which it will fit. The steps of this approximation algorithm are shown in Fig. 6.3. A straightforward implementation of the algorithm Fig. 6.3 runs in $O(n^2)$ time. However, a more sophisticated implementation reduces the running time to $O(n \log n)$ [54].

We note that the MM-PP problem can be reduced to the BIN PACKING problem as follows. Let P denote the given bound on peak power. For each action a_i with power value p_i, we create an item with size $= p_i/P$. Since $p_i \le P$, the size of each item is at most 1. Now, each time slot can be thought of as a bin (of unit capacity) so

---

1.Sort the items into non-increasing order of their sizes. Without loss of generality, let
   $\langle c1, c2, ..., cn \rangle$ denote the resulting sorted order.
2. **Comment:** Assign each item to the first bin in which it will fit.

    (a) Initialization: Let $j = 1$. (The variable j denotes the number of bins.) Let B_1 denote the
    initial bin.

    (b) **for** $i = 1$ **to** n **do**
       (i) Find the first bin, say bin B_k, among B_1 through B_j in which item i will fit.
       (ii) If there is no such bin, increment j by 1 and create a new bin B_j. Let $k = j$.
       (iii) Add item i to bin B_k.

3. Output j (the number of bins) and the packing generated above.

Fig. 6.3 Steps of the first fit decreasing algorithm for the BIN PACKING problem

that any packing of the items into q bins corresponds to a valid schedule of length q
and vice versa. Thus, any approximation algorithm for the BIN PACKING problem
can be used as an approximation algorithm for the MM-PP problem with the same
performance guarantee. A formal statement of this result is as follows.

Observation 6.4 *Any ρ-approximation algorithm for the* BIN PACKING *problem is
also a ρ-approximation algorithm for the* MM-PP *problem.*   ■

From the above discussion, we can conclude that there are efficient approximation algorithms with good performance guarantees for the MM-PP problem.

Application to Bluespec

In *Bluespec*, if the execution of a large set of non-conflicting actions in a single time
slot leads to the violation of peak power constraint, then such a set of actions can be
re-scheduled to execute over multiple time slots. Instead of executing all the actions
of the set, some of the actions can be postponed to execute in the future time slots in
order to meet the peak power constraint. Such a re-scheduling problem of *Bluespec*
is equivalent to the MM-PP problem, and hence the approximation algorithms for
the MM-PP problem can be used to re-schedule a set of actions of a *Bluespec* design
to meet the peak power constraint.

Note that executing only a subset of actions, say $A' \subset A$ in the present time slot
may disable (guards evaluate to *False*) the other actions in A which are scheduled
to execute in the future time slots. This may happen if the state elements updated
by the actions in A' executing in the present time slot are accessed by the guards
of the other actions in A which are postponed to future time slots. In such cases,
using the approximation algorithms of the MM-PP problem for re-scheduling of
the actions may result in changing the output of the design (though the output will
still be functionally correct) as compared to the original design [107]. However, for
designs described in terms of confluent set of actions, such change in the output
will not occur. This is because, as mentioned earlier, a confluent set of actions can
be executed in any order without changing the final state of the design. Thus, for a
design described using confluent set of actions, solutions of the MM-PP problem

can be used to re-schedule a set of non-conflicting actions to arrive at a minimum length schedule under the peak power constraints.

6.3.5 Approximation Algorithms for MPP-M

In the MPP-M problem, we are given a bound on the schedule length and the goal is to minimize the maximum power used in any time slot. Interestingly, when the power needed to execute each instruction is an integer, the MPP-M problem can be transformed into a classical multiprocessor scheduling problem [54]. In that problem, we are given a collection $T = \{T_1, T_2, \ldots, T_n\}$ of tasks, where each task T_i has an integer execution time e_i. The tasks are independent; that is, there are no precedence constraints among the tasks. The problem is to schedule the tasks in a non-preemptive fashion on m identical processors so as to minimize the makespan. To see the correspondence between the MPP-M problem and the classical scheduling problem, we think of each action a_i as a task and the power value p_i as the execution time of the corresponding task. Further, we think of the number of time slots ℓ as the number of available processors. Now, for the resulting scheduling problem, it can be seen that any schedule with makespan P on ℓ processors corresponds to a solution to the MPP-M problem using ℓ time slots and a peak power value of P. (The set of tasks scheduled to run on the same processor corresponds to the set of actions to be performed during the same time slot.)

In view of the above relationship between the MPP-M problem and the multiprocessor scheduling problem, any ρ-approximation algorithm for the latter is a ρ-approximation algorithm for the former. In particular, the following is a 4/3-approximation algorithm for the multiprocessor scheduling problem [58].

1. Construct a list of the tasks in non-increasing order of execution times.
2. Whenever a processor becomes available, assign the next job from the list to that processor. (If several processors become available at the same time, ties are broken arbitrarily.)

In terms of the MPP-M problem, the above approximation algorithm corresponds to sorting the actions in non-increasing order of their power requirements and assigning each action to a time slot for which the total power used is the smallest at that time. Clearly, this approximation algorithm can be implemented to run in $O(n \log n)$ time.

A PTAS is also known for the multiprocessor scheduling problem [58]. However, as the running time of the corresponding algorithm is exponential in $1/\varepsilon$, where $\varepsilon > 0$ is the chosen accuracy parameter, this may not be suitable in practice.

Application to Bluespec

For some *Bluespec* designs, latency is of prime concern. For such designs, their actions can be re-scheduled using the approximation algorithms for MPP-M problem in order to minimize the peak power of the design under the given bound on the latency. Similar to the MM-PP problem, here also the re-scheduling of the actions of a design may result in the disabling of some future actions, thus changing the output of the design. As discussed earlier, for designs described in terms of confluent set of actions, such re-scheduling is suitable (since it will not change their output) and hence can be used to minimize the peak power. Also, such re-scheduling is suitable for a persistent set of actions since executing a subset of such actions will not disable the other actions of the set.

Chapter 7
Dynamic Power Optimizations

Dynamic power is an important component of the power consumption of a hardware design. In this chapter, we present two algorithms that target the reduction of dynamic power during the CAOS-based synthesis process and produce RTL that can be synthesized into power-efficient hardware. We also present experimental results to show that when a CAOS specification is compiled using these algorithms, the resulting hardware (without any additional gate-level power optimizations) has power/area/latency numbers comparable to those obtained by using the well-known industrial-strength power optimization RTL to gate-level synthesis tools such as *Magma Blast Power* or *Synopsys Power Compiler*.

The experiments show that in the absence of such gate-level power optimization tools, the proposed algorithms show significant power reduction over standard *Bluespec Compiler (BSC)* for CAOS to RTL generation. Moreover, using the proposed algorithms, effects of various techniques on the power consumption of a design can be analyzed earlier in the design cycle (at RTL). This aids in faster architectural exploration by avoiding the need to go through the whole power estimation flow (up to gate-level) for each architectural choice.

7.1 Related Background

7.1.1 Clock-Gating of Registers

A large amount of dynamic power is consumed in the registers and the clocks of a design. Clock-gating of registers is a commonly used technique to reduce such power. To this end, Section 7.2 presents an algorithm that exploits the CAOS model of computation for efficient generation of gated-clocks during synthesis. This algorithm is general enough to apply to multiple clock domain specifications as well.

7.1.2 Operand Isolation

Another well-known technique to minimize the dynamic power of a hardware design is to decrease the switching activity at the inputs of its functional units.

G. Singh, S.K. Shukla, *Low Power Hardware Synthesis from Concurrent Action-Oriented Specifications*, DOI 10.1007/978-1-4419-6481-6_7, © Springer Science+Business Media, LLC 2010

During high-level synthesis, such decrease in the switching activity can be targeted during the scheduling, allocation, or binding phases of the synthesis process. However, even when the switching activity of various signals within a design is at its minimum, there is always a possibility of some unnecessary computation occurring in the design which can lead to unwanted power dissipation. A combinational computation is deemed unnecessary for a clock cycle if its output is not used for any useful purposes in that clock cycle.

Example

The CAOS-based description of GCD (Greatest Common Divisor) design can be written in terms of actions *Swap* and *Diff* as shown in Fig. 7.1. g_1 and g_2 are the guards of actions *Swap* and *Diff* respectively (x and y are the registers). The swap of the values in the body of action *Swap* occurs only when g_1 evaluates to *True*. The subtraction operation $y - x$ in the body of action *Diff* occurs whenever the values of x and/or y change but the assignment $y <= y - x$ occurs only when g_2 evaluates to *True*. Hence, when g_2 evaluates to *False*, the combinational logic corresponding to the subtraction operation is involved in unnecessary computation.

$$\textbf{Action Swap} : \mathbf{g_1} \equiv ((\mathbf{x} > \mathbf{y}) \ \&\& \ (\mathbf{y} \neq \mathbf{0}))$$
$$\mathbf{x} <= \mathbf{y};$$
$$\mathbf{y} <= \mathbf{x};$$

$$\textbf{Action Diff} : \mathbf{g_2} \equiv ((\mathbf{x} \leq \mathbf{y}) \ \&\& \ (\mathbf{y} \neq \mathbf{0}))$$
$$\mathbf{y} <= \mathbf{y} - \mathbf{x};$$

Fig. 7.1 CAOS description of GCD design

As illustrated in the above example, for a CAOS-based design computations occurring in the bodies of the actions whose guards evaluate to *False* should be avoided for the purposes of power savings. So far synthesis engines such as *BSC* [18] do not exploit this. An algorithm for efficiently blocking the switching activities to such unused parts of a design is presented in Section 7.3.

7.2 Clock-Gating of Registers

As already mentioned, technique of clock-gating of registers is commonly used at the RTL and lower levels of abstraction for reducing the register/clock power of a design. Let us consider a design described using guarded atomic actions with the following notations:

S: Set of all the state elements of the design.

A: Set of all the *actions* of the design.

g_i: Expression denoting the guard of an action $a_i \in A$.
R: $\{r : r \text{ is a register of the design}\}$, $R \subseteq S$.
U_r: Set of actions updating register r, $U_r \subseteq A$.
Clk_r: Clock of register r.
Rs_r: Reset signal of register r.
En_r: Enable of register r; that is, $En_r = (\bigvee_i g_i)$, $a_i \in U_r$.

$gatedClock(clk, en, rst)$: Function which returns a gated-clock (generated using a latch and an AND gate). It takes clock clk, enable signal en, and reset signal rst as the inputs.

C: Set of subsets of R where each subset contains registers having same clock; that is,
$$C = \{C_i : C_i \subseteq R; R = \bigcup_i C_i; C_i \cap C_j = \phi \,\forall\, i,j, i \neq j;$$
$$Clk_{r_a} == Clk_{r_b} \,\forall\, r_a, r_b \in C_i;$$
$$Clk_{r_x} \neq Clk_{r_y}, r_x \in C_i, r_y \in C_j \,\forall\, x, y, i, j, i \neq j\}$$

E: Set of subsets of $C_i \in C$ where each subset contains registers having same enable signal; that is,
$$E = \{E_j : E_j \subseteq C_i; C_i = \bigcup_j E_j; E_j \cap E_k = \phi \forall j, k, j \neq k;$$
$$En_{r_a} == En_{r_b} \,\forall\, r_a, r_b \in E_j;$$
$$En_{r_x} \neq En_{r_y}, r_x \in E_j, r_y \in E_k \,\forall x, y, j, k, j \neq k\}$$

T: Set of subsets of $E_j \in E$ where each subset contains registers having same reset; that is,
$$T = \{T_k : T_k \subseteq E_j; E_j = \bigcup_k T_k; T_k \cap T_l = \phi \,\forall\, k, l, k \neq l;$$
$$Rs_{r_a} == Rs_{r_b} \,\forall\, r_a, r_b \in T_k;$$
$$Rs_{r_x} \neq Rs_{r_y}, r_x \in T_k, r_y \in T_l \,\forall\, x, y, k, l, k \neq l\}$$

Algorithm 1 performs efficient clock-gating of registers for CAOS-based designs. In CAOS, registers which are updated by the same set of actions will have the same enable signal. This implies that an enable signal of a register is a disjunction of guards of all the actions which can update it. During clock-gating of registers, same gated-clock can be passed to registers having common enable signal (assuming same clock and reset signal). Thus, in CAOS-based designs, guards of the actions provide an efficient way of selecting which registers should share the gated-clocks.

Based on this idea, Algorithm 1 generates and assigns gated-clock to each register of a design. It efficiently handles designs with multiple clocks and reset signals. In designs with single clock and single reset signals (most designs fall in this category), gated-clocks will be assigned to the registers based on their enable signals, and hence the total number of generated gated-clocks will be equal to the number of distinct enable signals.

Algorithm 1 – Automatic Clock-Gating of Registers

---

for all r such that $r \in R$ **do**
    Compute enable signal En_r;
end for
Compute C = $\{C_i : C_i$ is a group of registers having same clock CLK$\}$;
for all C_i such that $C_i \in C$ **do**
    Compute set E = $\{E_j : E_j$ is a group of registers having same enable signal
EN$\}$;
    for all E_j such that $E_j \in E$ **do**
        Compute set T = $\{T_k : T_k$ is a group of registers having same reset signal
RS$\}$;
        for all T_k such that $T_k \in T$ **do**
            $gCLK = gatedClock(CLK, EN, RS)$;
            Replace clocks of all the registers in T_k by gated clock $gCLK$;
        end for
    end for
end for

7.3 Insertion of Gating Logic

In CAOS, the values of various guards of the design are computed in each clock cycle and are used to select the actions which can be executed in that clock cycle. Thus the combinational logic corresponding to a guard g_i is involved in useful computation in every clock cycle. On the other hand, expressions in the bodies of various actions also compute values in each clock cycle but only some of those values (having corresponding g_i evaluate to *True*) are selected to update the state of the design. The computations corresponding to the unselected values can be considered as *unnecessary computations* since those values are not used to compute the next state. Avoiding such computations will result in reduction of the switching activity of the design leading to dynamic power savings.

As mentioned earlier, the guards and the bodies of the actions are composed of one or more expressions. Algorithm 2 parses the expressions corresponding to the bodies of various actions and inserts gating logic (using AND gate or LATCH) at the appropriate places in these expressions. Each expression gets translated into a combinational logic during synthesis. The gating logic is inserted such that the inputs of these combinational blocks are isolated/gated using guard of the corresponding action as the activation signal of the gate. Thus computations are triggered across a combinational logic only when its output is used in some further computations or to update the state of the design. For efficient gating of various signals of a design, two problems need to be solved:

1. *Insertion of gates at the appropriate points*– Algorithm 2 targets this problem by parsing through various expressions used in the bodies of the actions of a design and inserting gates such that the unnecessary computations are minimized. While inserting the gates, expressions used in the guards are not affected since outputs of such expressions are involved in useful computation in each clock cycle. Sharing of common expressions among various actions is also taken into account while inserting the gating logic.
2. *Selection of activation signal* – The guard of each action can be used to decide if the computation occurring within a combinational logic will be used in a clock cycle. Thus, for a design described using CAOS, activation signals required for the gating logic are already exposed in the form of these guards, which makes the implementation of gating logic efficient since no separate circuit is required for the generation of these activation signals.

Consider the following notations:

G: Set of expressions corresponding to guards of the actions of the design ($g_i \in G$).

B_i: Set of expressions (including the ones involved in the composition of other expressions) used in the body of an action $a_i \in A$.

E_g: Set of expressions (including the ones involved in the composition of other expressions) used in various guards of the design.

For the GCD design shown in Fig. 7.1, guards g_1 and g_2 are expressions which can be expressed in terms of other expressions as shown in Fig. 7.2.

$$e_4 = x.\text{read}(); \quad e_5 = y.\text{read}();$$
$$e_1 = e_4 > e_5; \quad e_2 = e_4 \le e_5; \quad e_3 = e_5 \ne 0;$$
$$g_1 = e_1 \,\&\&\, e_3; \quad g_2 = e_2 \,\&\&\, e_3;$$

Fig. 7.2 Expressions used in GCD design

An expression can denote complicated operations on the state of a design (in which case it can be composed of one or more other expressions) or it can be as simple as reading the value of a state element, for example, $e_4 = x.\text{read}()$ as shown in Fig. 7.2. Thus, for the GCD design, we have

$S = \{x, y\}; \quad G = \{g_1, g_2\};$
$A = \{a_1, a_2\}$ where a_1 denotes action *Swap* and a_2 denotes action *Diff*;
$B_1 = \{e_4, e_5\}; \quad B_2 = \{e_6, e_4, e_5\}$ where $e_6 = e_5 - e_4;$
$E_g = \{e_1, e_2, e_3, e_4, e_5\};$

Let us define the following functions:

subExprs(e): Function that returns the set of expressions that are used to compose expression e. Such expressions can also be considered as the inputs to expression e.

isValueRead(e): Function that returns *True* if expression *e* represents an access (reading a value) to a memory element.

rank(e): Function that returns the number of actions that share an expression *e*. It can be defined as

$$\text{rank}(e) = \sum_{i=1\text{to}|A|} n_i \qquad (7.1)$$

where $n_i = 1$ if $e \in B_i$, else $n_i = 0$

gate(e, g_i): Function that inserts the gating logic. It returns a new expression e' which evaluates to *e* when g_i is *True* or evaluates to zero otherwise. Such an expression can be composed using any of the following definitions:

1. *Using AND gate ($e' = e$ && g_i)* – Gating using AND gates is mainly suitable for designs where guard g_i does not change frequently; that is, actions of the design do not execute frequently. This is because an AND gate will change its output when g_i transitions from high to low, thus triggering some unnecessary computation in the combinational logic during such a transition.
2. *Using LATCH ($e' = $ Output of a latch having input as e and enable as g_i)* – Gating using latches is suitable for designs where actions execute frequently. This is because a latch will hold the output value if its enable signal is low and no transitions occurs at their outputs when g_i changes from high to low.

isolate(e, i): Function that returns a new expression after inserting appropriate gating logic in expression *e*. It uses guard of action a_i as the activation signal for the gating logic.

The algorithm starts by selecting an action a_i and parsing through each expression used in its body. For each such expression *e* the algorithm makes call to function *isolate(e,i)* which handles the following cases:

1. *Expression e is also used in composing at least one guard ($e \in E_g$)* – In CAOS, each guard $g_i \in G$ is involved in useful computation in every clock cycle. Since $e \in E_g$, its output will also be used in each clock cycle.

Algorithm 2 – Inserting Gates for Power Savings

$E_g = allGuardExprs(G)$;
for all a_i such that $a_i \in A$ **do**
    for all *e* such that $e \in B_i$ **do**
        $isolate(e, i)$;
    end for
end for
FUNCTION isolate (e, i) //Inserts gating logic into expression *e*.
if $(e \in E_g)$ **then**
    $e' = gate(e, g_i)$;   replace e by e' in B_i;   return;

else if $(rank(e) == 1)$ **then**
  if $(isValueRead(e) == True)$ **then**
    $e' = gate(e, g_i)$;   replace e by e' in B_i;   return;
  else
    for all e' such that $e' \in subExprs(e)$ **do**
      $isolate(e', i)$;
    end for
  end if
else
  if $(isValueRead(e) == True)$ **then**
    $e' = gate(e, g_i)$;   replace e by e' in B_i;   return;
  else
    $e' = e$;   replace e by e' in B_i; $isolate(e', i)$;
  end if
end if
FUNCTION allGuardExprs(G) //Computes all expressions used in guards.
$R = G$;
for all g_i such that $g_i \in G$ **do**
  $E = allSubExprs(g_i)$; $R = R \cup E$;
end for
return R;
FUNCTION allSubExprs(e) //Computes expressions composing e.
$S = subExprs(e)$;
if $S == NULL$ **then**
  return $NULL$;
else
  $T = S$;
  for all e such that $e \in S$ **do**
    $X = allSubExprs(e)$; $T = T \bigcup X$;
  end for
  return T;
end if

Thus, no further parsing of e is required; that is, the expressions which are used to compose e need not be parsed. In this case, a new expression $e' = gate(e, g_i)$ which incorporates the gating logic is created. Then, e is replaced by e' in the body B_i of the selected action a_i. This makes sure that the guards of the actions are not affected by the insertion of the gating logic since they will use the output of expression e for proper evaluation.

2. *Expression e is used only in the body of* a_i $(rank(e) == 1)$ – If $isValueRead(e)$ returns *True*, then such an expression can be gated for power savings (without further parsing the expressions used in its composition). So a new expression e' which incorporates the gating logic is created and e is replaced by e' in B_i. On the other hand, if $isValueRead(e)$ returns *False*, then the expressions used for

the composition of e are parsed further for the insertion of gating logic at the appropriate points.

3. *Expression e is used by the body of at least one more action* ($rank(e) > 1$) – If *isValueRead(e)* returns *True*, then a new expression e' which incorporates the gating logic is created and e is replaced by e' in B_i. This way, the other action(s) which are also using expression e will not be affected. On the other hand, if *isValueRead(e)* returns *False*, then a new expression $e' = e$ is created. The idea is to avoid sharing and create a separate combinational logic for the body of the selected action a_i which can be gated independently without affecting other actions. In most cases, such duplication of combinational logic increases the potential of gating these blocks independent of each other, thus leading to higher power savings. After creating e', expression e is replaced by e' in B_i and expressions which are used to compose expression e' are parsed further.

7.3.1 Other Versions of Algorithm 2

7.3.1.1 Version 2

Case 2 of Algorithm 2 occurs when expression e is used in the body of only one action; that is, $rank(e) == 1$. As mentioned earlier, if *isValueRead(e)* returns *True*, then such an expression e is gated without any further parsing of the expressions used in its composition. A different implementation of Algorithm 2 can be obtained by gating e and then continue parsing the expressions used in its composition in order to look for more opportunities of isolation. Such opportunities of isolation may arise in cases when a value is read from a memory element based on the value of some other argument, and hence the expressions corresponding to both these values can be isolated.

7.3.1.2 Version 3

Another possible implementation of Algorithm 2 can be obtained by modifying its Case 3. As mentioned earlier, Case 3 occurs when expression e is used by the body of at least one more action; that is, $rank(e) > 1$. Let A' be the set of all such actions using e. Let G' be the set of expressions corresponding to the guards of all the actions in A'.

In Case 3, expression e' is used to replace expression e as part of isolation. When *isValueRead(e)* returns *True*, e' is computed as $e' = gate(e, g_i)$, otherwise $e' = e$. Instead, in both these cases, expression e' can be evaluated using a new activation signal a which can be given as

$$a = \vert\vert_{i=1 \text{to} |G'|} \ g_i, \text{ where } g_i \in G' \tag{7.2}$$

Thus, instead of using the guard of an action as an activation signal to evaluate e', the disjunction of all the expressions in G' (corresponding to the guards of all the actions using expression e) can be used as the activation signal. For example, in case AND gates are used for isolation, expression e' can be created as $e' = e$ && a.

7.4 Experiment and Results

We implemented the proposed algorithms in *Bluespec Compiler* (*BSC*) and tested them on various realistic designs. The selected designs vary in their nature, size, and complexity. Each design is first synthesized to RTL using *BSC* which converts a CAOS-based description of a design into RTL Verilog code. Logic synthesis tools like *Magma's Blast Create* (*version 2005.03.127-linux24_x86*) and *Synopsys' Power Compiler* (*version Y-2006.06-SP2 for Linux*) are then used to convert the RTL description of the design to gate-level netlist using the $0.18\,\mu$m TSMC technology library. In order to compare the performance of the algorithms under different tools, we present results corresponding to synthesis from *Blast Create* as well as *Power Compiler*. The generated RTL and gate-level Verilog codes are simulated using *Synopsys VCSi* (*version X-2005.06-9*) to verify the functional behavior of the designs. The synthesized designs are checked to make sure that they meet the timing requirements. Power estimation is done at both RTL and gate-level for which the generated Verilog design files and the simulation activity files (in value change dump (vcd) format) are passed to *Sequence PowerTheater* (*version R2006.1*) [99]. Both RTL and gate-level experimental results for Algorithms 1 and 2 are presented below.

7.4.1 Algorithm 1

Dynamic power of a design is composed of its *Combinational Power, Register Power*, and *Clock Power*. Tables 7.1 and 7.3 show the reductions (as fractional change from original power) obtained in *Total Power* and *Register+Clock Power* using Algorithm 1. The numbers shown are obtained by performing power estimation at the gate level.

Table 7.1 shows the results obtained when *Blast Create* is used as the logic synthesis tool. Since clock-gating of registers can also be handled efficiently by logic synthesis tools, in Table 7.1 we compare the power savings achieved using

Table 7.1 Power savings using Algorithm 1 compared with Blast Create's results

Design	Original power (mW) (total/register+clock)	Power after Blast Create's clock-gating (fractional change) (total/register+clock)	Power after using Algorithm 1 (fractional change) (total/register+clock)
AES	111.00/42.30	0.66/0.28	0.69/0.21
DMA	20.40/17.79	0.55/0.54	0.65/0.65
GCD	4.26/2.73	0.71/0.63	0.69/0.58
UC	55.50/50.70	0.90/0.92	0.91/0.90
FSM	33.50/7.62	0.93/0.70	0.92/0.66
VM	0.46/0.25	0.80/0.60	0.79/0.66

Algorithm 1 against the savings obtained by turning on *Blast Create's* clock-gating feature.

Insertion of the extra clock-gating circuitry for power savings of a design is associated with corresponding increase in its area. Table 7.2 shows the area penalties (as fractional change relative to area of the original design) reported by *Blast Create* on using Algorithm 1 for power savings. Corresponding power and area numbers when *Power Compiler* is used as the logic synthesis tool are shown in Tables 7.3 and 7.4.

Table 7.2 Area penalties using Algorithm 1 compared with Blast Create's results

Design	Original area (μm^2)	Area after Blast Create's clock-gating (fractional change)	Area after using Algorithm 1 (fractional change)
AES	627, 572	0.94	0.94
DMA	90, 422	0.81	0.89
GCD	18, 262	0.91	0.91
UC	233, 739	0.81	0.98
FSM	154, 415	0.98	0.98
VM	1, 873	0.99	1.00

Table 7.3 Power savings using Algorithm 1 compared with Power Compiler's results

Design	Original power (mW) (total/register+clock)	Power after Power Compiler's Clock-gating (fractional change) (total/register+clock)	Power after using Algorithm 1 (fractional change) (total/register+clock)
AES	74.60/41.30	0.50/0.16	0.55/0.17
DMA	18.20/16.18	0.59/0.57	0.63/0.62
GCD	4.51/2.73	0.55/0.38	0.61/0.48
UC	48.00/44.40	0.89/0.89	0.88/0.88
FSM	22.30/6.31	0.87/0.54	0.84/0.58
VM	0.41/0.25	1.04/0.86	0.90/0.64

Table 7.4 Area penalties using Algorithm 1 compared with Power Compiler's results

Design	Original area (μm^2)	Area after Power Compiler's clock-gating (fractional change)	Area after using Algorithm 1 (fractional change)
AES	586, 481	0.93	0.92
DMA	85, 166	0.84	0.83
GCD	18, 448	0.89	0.89
UC	238, 613	0.97	0.97
FSM	125, 818	0.97	0.97
VM	1, 883	0.96	0.97

As shown in Tables 7.1 and 7.3, Algorithm 1 consistently showed significant power savings in all the designs. Larger power savings were obtained for *AES* and *DMA* designs which consist of several registers that are not updated frequently, thus saving significant power by clock-gating of registers.

Comparison of the results shows that Algorithm 1 is competitive in the sense that for most designs power saved by using Algorithm 1 is very close to the savings achieved by using *Blast Create's* or *Power Compiler's* clock-gating. For some designs like *Vending Machine* (*VM*) (Tables 7.1 and 7.3), *Greatest Common Divisor* (*GCD*) (Table 7.1), and *FSM* (Table 7.3), Algorithm 1 even performs better than the logic synthesis tools. This can be attributed to the fact that CAOS, which is at a higher level of abstraction than RTL, can facilitate in taking efficient decisions during the application of low-power techniques.

7.4.2 Algorithm 2

Tables 7.5, 7.6, 7.7, and 7.8 show the reductions obtained in *Total Power* and *Combinational Power* of various designs using Algorithm 2 (and its versions) along with the associated effects on the area of those designs. The reported numbers (gate-level) are obtained with AND gates used as the gating logic in order to minimize the power and area overheads associated with the extra circuit inserted by Algorithm 2. For each design, all power and area results are shown as fractional change as compared to the original design.

7.4.2.1 Logic Synthesis Using Blast Create

In Table 7.5 we show the power savings achieved using Algorithm 2 and its versions when *Blast Create* is used for logic synthesis. Total power savings of up to 25% (*AES* design) on using Version 2 of the algorithm demonstrates that Algorithm 2 can be successfully used to generate power-efficient designs. *AES* design, which is an implementation of the *Advanced Encryption Standard* (*AES*) algorithm, consists of 11 actions only some of which were executing in each clock cycle. Thus, Algorithm 2 showed significant power savings for *AES* design. Other designs like *DMA*, *FSM*, *Vending Machine* (*VM*) also show a decrease in the total power consumption on using Algorithm 2 as shown in Table 7.5. Note that in most cases, Version 1 of Algorithm 2 shows maximum power savings. But for the *AES* design, Version 2 performs even better than Version 1. This is because, as explained earlier, Version 2 of the algorithm looks for extra opportunities of isolation by further tracing the expressions involved in accessing the value of a memory element. Thus, depending on the design either of these versions can be used for power savings.

On the other hand, as shown in Table 7.5, for *UC* design (an implementation of Bus Upsize Converter) we noticed an increase in its power consumption on using Algorithm 2. Further analysis showed that for designs in which most actions execute frequently, using Algorithm 2 may increase their power demand. This is because the inserted gating logic also consumes some additional power and if the combinational power saved is less (due to frequent execution of most combinational logic) than this extra overhead, then the overall power of the design will increase. We noticed that for most CAOS-based designs, an action which can be successfully gated (an action is said to be successfully gated in a clock cycle if its guard evaluates to *False* in that cycle) for more than two consecutive clock cycles will contribute to power savings on using Algorithm 2. On the other hand, an action which executes

Table 7.5 Power savings using Algorithm 2 and its versions synthesized using Blast Create

Design	Original power (mW) (total/combinational)	Power using Algorithm 2 (fractional change) (total/combinational)	Power using Algorithm 2 – Version 2 (fractional change) (total/combinational)	Power using Algorithm 2 – Version 3 (fractional change) (total/combinational)
AES	111.00/68.50	0.85/0.76	0.75/0.58	0.88/0.79
DMA	20.40/2.64	0.95/1.13	0.95/1.19	0.95/1.21
GCD	4.26/1.53	0.99/1.02	0.99/1.02	0.99/1.02
UC	55.50/4.77	1.01/1.15	1.03/1.32	1.03/1.21
FSM	33.50/25.90	0.92/0.91	0.92/0.91	1.35/1.46
Vending Machine (VM)	0.46/0.21	0.98/1.03	0.98/1.03	0.98/1.03

Table 7.6 Area penalties using Algorithm 2 and its versions synthesized using Blast Create

Design	Original area (μm^2)	Area after Algorithm 2 (fractional change)	Area after Algorithm 2 – Version 2 (fractional change)	Area after Algorithm 2 – Version 3 (fractional change)
AES	627,572	1.06	1.01	0.98
DMA	90,422	0.98	0.99	0.98
GCD	18,262	1.06	1.06	1.06
UC	233,739	1.01	1.01	1.01
FSM	154,415	1.39	1.39	1.06
VM	1,873	1.12	1.12	1.12

frequently (and thus cannot be gated for large number of consecutive cycles) may result in an increase in the design's power consumption.

Power savings obtained by using Algorithm 2 are also associated with a corresponding increase in the area of a design due to the insertion of extra gating logic. Table 7.6 reports the associated area penalties for each design on using Algorithm 2 (the numbers are obtained from the area reports generated by *Blast Create*). Maximum area penalties are seen for the *FSM* design. Since the chosen *FSM* consisted of a large combinational part, using Algorithm 2 for power savings resulted in the insertion of significant gating logic, thus increasing the area of the design. Hence, application of Algorithm 2 involves a power-area trade-off.

Addition of extra gating logic for the purposes of power savings also affects the critical path slack for a design, thus affecting its performance. This is because the computation corresponding to the guard and the body of an action are forced to occur sequentially (as opposed to concurrent execution in the original design) due to the added gating logic. Thus, on using Algorithm 2 the slack of a design should usually shrink. However, we noticed that for some designs Algorithm 2 actually resulted in some slack improvement. This can be attributed to the fact that the addition of extra AND gates enables some additional Boolean optimizations during logic synthesis of these designs. Such optimizations may also result in slight area reduction in some cases.

7.4.2.2 Logic Synthesis Using Power Compiler

Tables 7.7 and 7.8 show the corresponding power and area numbers when *Power Compiler* is used for logic synthesis. There, we also compare the results obtained by using Algorithm 2 (and its versions) against the power reductions obtained using operand-isolation feature of *Power Compiler*. (We could not do such a comparison of results for *Blast Create* since it does not support operand isolation and we do not have access to *Magma's Blast Power* tool which supports this feature.) For operand isolation by *Power Compiler*, the two-pass approach (based on incremental compilation of the design) as recommended in the *Power Compiler* user guide is used.

As shown in Table 7.7, power savings achieved using Algorithm 2 and its versions are comparable to the savings obtained using *Power Compiler's* operand isolation. Version 2 of the algorithm saves maximum power for the *AES* design, whereas

Table 7.7 Power savings using Algorithm 2 and its versions synthesized using Power Compiler

Design	Original power (mW) (total/combinational)	Power using PC Op. Iso. (fractional change) (total/combinational)	Power using Algorithm 2 (fractional change) (total/combinational)	Power using Algorithm 2 – Version 2 (fractional change) (total/combinational)	Power using Algorithm 2 – Version 3 (fractional change) (total/combinational)
AES	74.60/33.30	0.99/0.97	1.00/1.01	0.88/0.73	1.01/1.03
DMA	18.20/2.06	1.01/1.03	1.01/1.02	1.01/1.02	1.02/1.09
GCD	4.51/1.79	0.99/1.00	1.18/1.34	1.18/1.33	1.18/1.33
UC	48.00/3.52	1.00/1.03	1.00/0.99	1.01/1.05	1.00/0.98
FSM	22.30/16.00	0.87/0.78	0.83/0.76	1.03/1.01	1.26/1.34
Vending Machine (VM)	0.41/0.15	1.12/1.40	1.13/1.28	1.13/1.28	1.13 /1.28

Table 7.8 Area penalties using Algorithm 2 and its versions synthesized using Power Compiler

Design	Original area (μm^2)	Area after PC Op. Iso. (fractional change)	Area after Algorithm 2 (fractional change)	Area after Algorithm 2 – Version 2 (fractional change)	Area after Algorithm 2 – Version 3 (fractional change)
AES	586, 481	1.00	1.07	1.07	0.99
DMA	85, 166	1.02	1.00	1.00	1.00
GCD	18, 448	1.00	1.25	1.25	1.25
UC	238, 613	1.00	1.00	1.02	1.00
FSM	125, 818	1.16	1.46	1.46	1.07
VM	1, 883	1.24	1.17	1.17	1.17

for the *FSM* design the original Algorithm 2 results in most power savings. Version 3 performed slightly better than others in case of *UC* design. The results show that for most designs either Algorithm 2 or its Version 2 provides better power savings. Thus, depending on the design appropriate version of Algorithm 2 can be used.

Note that, as shown in Table 7.7, the power consumption of some designs remains almost the same as the original design on using Algorithm 2 or *Power Compiler's* operand isolation. We need to develop a better understanding of *Power Compiler's* synthesis process to reason about this behavior.

7.4.2.3 Another Refinement

As mentioned earlier, in Algorithm 2 guards of various actions are used in the gating logic (as the activation signals) for isolating a part of the design. In the real hardware, any unnecessary switching activity occurring in the guards of a design (before the signal settles down to a value) will result in extra power consumption. In order to avoid the propagation of such switching occurrences in various guards, their values can be passed to the gating logic only at the negative edge of the clock.

We implemented such a refinement of Algorithm 2 using latches which pass the values of the guards to the gating logic only at the negative edge of the clock. We noticed that for some designs such a refinement helps to further decrease the power of the design at the cost of extra area overhead incurred by the addition of latches. Also, such a use of latches results in increasing the clock period of a design since the values of guards are only passed in the second half of the clock cycle, thus leaving less time for various computations to complete.

7.4.3 RTL Power Estimation

Instead of performing logic synthesis for analyzing the power consumed by a design, *Power Theater* can be used for RTL power estimation so that the affects of various low-power techniques can be evaluated earlier in the design cycle. This aids in faster architectural exploration. Tables 7.9 and 7.10 show the power savings achieved using Algorithms 1 and 2, respectively, when power estimation is done at the RTL.

Table 7.9 RTL power savings using Algorithm 1

Design	Original power (mW) (total/register+clock)	Power after using Algorithm 1 (fractional change) (total/register+clock)
AES	68.20/48.30	0.43/0.20
DMA	18.80/17.59	0.63/0.62
GCD	2.84/2.11	0.73/0.63
UC	54.10/51.30	0.90/0.89
FSM	20.40/6.32	0.91/0.74
VM	0.39/0.24	0.82/0.73

Table 7.10 RTL power savings using Algorithm 2 and its versions

Design	Original power (mW) (total/combinational)	Power using Algorithm 2 (fractional change) (total/combinational)	Power using Algorithm 2 – Version 2 (fractional change) (total/combinational)	Power using Algorithm 2 – Version 3 (fractional change) (total/combinational)
AES	68.20/19.80	0.98/0.95	0.95/0.83	0.99/0.97
DMA	18.80/1.21	1.03/1.33	1.03/1.31	1.01/1.14
GCD	2.84/0.73	0.96/0.85	0.96/0.84	0.96/0.85
UC	54.10/2.87	0.99/0.92	1.00/0.95	0.99/0.92
FSM	20.40/14.00	0.81/0.76	0.81/0.76	0.99/1.04
Vending Machine (VM)	0.39/0.14	0.91/0.78	0.91/0.78	0.91/0.78

Comparison of the RTL power numbers against the gate-level numbers shows that RTL power estimation can be successfully used to analyze the affects of Algorithms 1 and 2 on the power consumption of most designs. However, for some designs like *AES* (Table 7.1), *FSM* (Table 7.1), and *GCD* (Table 7.3) there were significant differences in the absolute RTL and gate-level power numbers. However, as shown in Tables 7.9 and 7.10, even for these designs the fractional power savings at RTL and gate-level are similar, thus supporting the fact the RTL power estimation can be successfully used for analyzing low-power techniques at a level above RTL.

Thus, applying such low-power techniques earlier in the design (above RTL) aids in faster exploration leading to a significant increase in the designer's productivity. Moreover, in some cases applying these techniques at a higher level of abstraction helps in taking efficient decisions to further increase the achieved power savings. In other words, the above results show that applying low-power techniques above RTL may aid in extra power savings in addition to offering the advantage of earlier assessment of the affects of these optimizations on the power consumption of a design.

7.5 Summary

In this chapter, we presented two algorithms for dynamic power reduction in hardware designs generated during CAOS-based synthesis process. Algorithm 1, presented in Section 7.2, provides an efficient method for the generation of appropriate gated-clocks for registers of a design. Such assignment of gated-clocks becomes difficult, and hence may be inefficient at lower levels of abstraction. Section 7.3 presents Algorithm 2 which performs automatic insertion of gating logic for combinational power reduction. It exploits the fact that if an action is not executed in the present clock cycle, then the computation occurring in its body can be avoided for the purposes of power savings. However, application of Algorithm 2 is associated with the following issues:

1. Algorithm 2 may increase the power consumption for designs where most actions are executing frequently.
2. Power savings obtained using Algorithm 2 are associated with a corresponding increase in the area of the design.
3. Inserting gating logic in critical paths of a design might not be desirable because it can lead to timing issues.

These issues can be resolved by refining Algorithm 2. Instead of inserting the gating logic in all the actions of a design, selective gating can be done. For example, gating logic can be implemented only in those actions of the design that do not execute frequently. Such refinements can be guided by feeding back the execution traces of the actions of a design to *BSC*. Moreover, the use of appropriate gates (AND gate or LATCH) depending on the nature of the design can further improve its power savings.

The presented experimental results demonstrate the effectiveness of using the proposed algorithms for dynamic power reduction of CAOS-based designs. As expected, the algorithms result in power savings for most designs at the cost of the corresponding area/timing penalties. The results obtained using these algorithms are comparable to those obtained after similar power optimizations are done using commercial logic synthesis tools. Hence, applying such power optimization techniques at CAOS level facilitates the earlier (above gate-level) assessment of the effects of such techniques on the power, area, and latency of a design, thus aiding in faster architectural exploration and enhanced productivity.

Chapter 8
Peak Power Optimizations

In hardware generated using CAOS, concurrent execution of maximal set of actions in each clock cycle reduces the latency (number of clock cycles) of the design at the cost of increase in its peak power, which is defined as the maximum instantaneous power (due to switching activity) during one clock cycle. In Chapter 5, various heuristics targeting the minimization of peak power in designs generated from CAOS are proposed. The peak power heuristics presented in that chapter postpone some actions (among all the actions that can be executed in a clock cycle) to the future clock cycles for peak power reduction. Also, actions which were postponed in the previous clock cycles are considered for execution in the present clock cycle. For this, those heuristics propose the use of extra state elements in order to remember which actions were postponed in the previous clock. The main drawback of those approaches is that such a use of extra state elements is associated with the corresponding area and power overheads.

The heuristic presented in this chapter eliminates the use of such state elements. This is done by exploiting the fact that for a CAOS-based design, any number of actions can be disabled (not allowed to execute) in a clock cycle for peak power reduction. Moreover, it is not necessary to remember which actions were disabled in the previous clock cycles. This is an artifact of using atomic actions-based models for describing hardware designs. In other words, a well-written design should allow disabling of any number of its actions in a clock cycle without affecting its functionality. A design should be written such that disabling some its enabled actions in a clock cycle automatically enables those actions in some future cycles, thus ensuring that the behavior of the design remains the same, albeit at the cost of increased latency. Thus, while writing a design using CAOS, a designer should not rely on the fact that during the execution of the design maximal set of actions will be executed in a clock cycle. Such well-written designs increase the scope of scheduling-based optimizations that can be performed during CAOS-based synthesis process. In this chapter, we propose an algorithm for reducing the peak power of such designs. During the execution, the proposed algorithm decides what actions of a design can be safely disabled in a clock cycle such that the degradation of the latency of the design is minimized while satisfying its peak power constraint.

We implemented the proposed algorithm in *BSC* and tested it using various realistic designs. Experimental results shown in Section 8.4 demonstrate that the

G. Singh, S.K. Shukla, *Low Power Hardware Synthesis from Concurrent Action-Oriented Specifications*, DOI 10.1007/978-1-4419-6481-6_8,

proposed algorithm can reduce the peak power of hardware designs (up to 20% reduction achieved) without significant degradation in their latency. Like the algorithms presented in the previous chapters, this algorithm also exploits the high abstraction level of CAOS to offer the benefits of easier power management (instead of re-scheduling each individual operation of a design, all operations corresponding to a particular action are considered together during re-scheduling) and faster architectural exploration, thus increasing the overall productivity of the designer.

8.1 Related Background

In hardware generated from CAOS, operations corresponding to any two actions can be scheduled to execute in the same clock cycle such that their atomicity is maintained. This means that the state resulting from concurrent execution of the operations of two or more actions in a clock cycle corresponds to the state obtained by the sequential execution of those atomic actions in a particular order. However, in real hardware, maintaining such atomicity among various actions within the same clock cycle may lead to complicated combinational circuit. To avoid this, a notion of conflict is introduced and two actions can be considered to be conflicting with each other if executing their operations in the same clock cycle is either too complicated or undesirable for pragmatic reasons (like short critical paths, write–write conflicts, easier hardware analysis). A simple example of such a conflict is two actions updating the same hardware register. In the synthesized hardware, such restrictions (conflicts) are enforced using small overhead logic such that two conflicting actions are never allowed to execute in the same clock cycle.

During the CAOS-based synthesis process, automatic scheduling of various operations of a design can be done based on such conflicts. For latency minimization, maximal set of non-conflicting actions can be scheduled to execute in each clock cycle. However, executing large number of operations in a single cycle leads to an increase in the peak power of the design. One way of minimizing the peak power is to introduce some extra conflicts among various actions of a design. This will reduce the number of operations executing in each clock cycle, thus reducing the peak power. However, this may adversely affect the latency of the design. This is because addition of extra conflicts will enforce that two conflicting actions will never execute in the same clock cycle even if their concurrent execution does not violate the peak power constraint in some cycles. This is undesirable from a latency point of view. The algorithm presented later in this chapter allows the designer to specify a peak power constraint for a design, which is then used to schedule its actions such that degradation to the latency of the design is minimized. This is done by executing maximal set of actions in each clock cycle under the given peak power constraint.

Figure 8.1 shows CAOS-based description of the hardware implementation of a *Vending Machine* design. Most of the description is self-explanatory. The design is described as a module composed of three parts:

```
module mkVending()
// count of money in the vending machine
Reg(Int(7)) count = 0;

// state bit that controls dispensing money back
Reg(Bool) moneyBack = False;

// wire to control dispense money output
PulseWire dispenseMoney;

// wire to control gum dispenser output
PulseWire gumControl;

// action that controls dispensing of money
action doDispenseMoney() if (moneyBack)
    let newCount = count - 10;
    count <= newCount;
    dispenseMoney.send();
    if (newCount == 0)
        moneyBack <= False;
end;

// action that controls dispensing of gum
action doDispenseGum() if (!moneyBack && count >= 50)
    count <= count - 50;
    gumControl.send();
end;

// input-handling methods
method tenCentIn() if (!moneyBack)
    count <= count + 10;
end;

method fiftyCentIn() if (!moneyBack)
    count <= count + 50;
end;

method moneyBackButton() if (!moneyBack)
    moneyBack <= True;
end;

// connect wires for money and gum outputs
method dispenseTenCents()
    return dispenseMoney;
end;

method dispenseGum()
    return gumControl;
end;

endmodule
```

Fig. 8.1 Vending machine design

1. *State elements, count and moneyBack*, that correspond to the registers of the design. The state elements of a design can be in the form of registers, FIFOs, or memories.
2. *Atomic actions, doDispenseMoney and doDispenseGum*, that perform computations and update the state elements of the design under certain conditions (called guards – shown with *if* clause). Action *doDispenseMoney* controls the dispensing of the money whereas *doDispenseGum* controls the dispensing of gum.
3. *Interface methods, tenCentIn, fiftyCentIn, moneyBackButton, dispenseTenCents,* and *dispenseGum*, that interact with an external module such as a testbench or

other hardware design. Interface methods also behave like atomic actions but their execution is controlled by the external module.

Actions *doDispenseMoney* and *doDispenseGum* are mutually exclusive, whereas actions *doDispenseGum, tenCentIn*, and *fiftyCentIn* conflict with each other since they all update register *count*. (*Vending Machine* design of Fig. 8.1 also contains some combinational wires which are used for controlling the outputs of the design.)

8.2 Formalization of Peak Power Problem

Let us define a relation $<_s$ among any two non-conflicting actions a_i and a_j, such that $a_i <_s a_j$ holds if concurrent execution of a_i and a_j in a clock cycle is equivalent to executing a_i in the present clock cycle followed by a_j in the next cycle in a sequential order. In the Vending Machine design, actions *doDispenseGum* and *moneyBackButton* can be executed concurrently since their concurrent execution is equivalent to the following sequential ordering: *doDispenseGum, moneyBackButton*. This can be denoted as *doDispenseGum* $<_s$ *moneyBackButton*.

In order to generate appropriate scheduling and control logic that maintains the atomicity of various actions, synthesis from atomic actions involves constructing (at compile time) a single sequential ordering S_{order} consisting of all the actions of a design. To arrive at such an ordering, transitivity property of relation $<_s$ is used and cycles are broken appropriately. For the Vending Machine design, one such possible ordering S_{order} of actions is given as

tenCentIn, fiftyCentIn, doDispenseMoney, dispenseTenCents, doDispenseGum, moneyBackButton, dispenseGum.

For CAOS-based designs, the correctness constraint (which requires the concurrent execution of multiple actions to correspond to at least one sequential ordering of their execution) actually provides flexibility in terms of re-scheduling some of the actions for peak power reduction. Based on the sequential ordering, only an appropriate subset of non-conflicting actions can be allowed to execute in each clock cycle in order to reduce the peak power of a design without altering its functional behavior. This subset can be chosen such that the degradation of the latency of the design is minimized. All other enabled actions should be disabled (not allowed to execute) in a clock cycle.

However, in some cases disabling an enabled action a_i from executing in a particular clock cycle may affect the enabling of some other action a_j in that cycle. This might happen due to a combinational path from a signal denoting disabling of a_i to a signal used to compute the guard of a_j. Let $<_d$ be the relation between two actions a_i and a_j denoting such a combinational path dependency. For example, in the Vending Machine design, such a relation exists between conflicting actions *tenCentIn* and *doDispenseGum* (denoted as *tenCentIn* $<_d$ *doDispenseGum*) since *tenCentIn* corresponds to a method call and disabling its execution in a clock cycle (for peak power reduction) affects the enabling of *doDispenseGum* through a combinational path.

Consider a design described using guarded atomic actions with the following notations:

  A: Set of all the actions used to describe a design.
  P: Peak power constraint of the design.
  p_i: Power consumed on executing action $a_i \in A$ in a clock cycle. It is also called power-weight of action a_i.

Peak Power Problem – Given a set $A_c \subseteq A$ containing maximal number of non-conflicting actions that are selected for execution in a clock cycle c, determine a set $A_c^P \subseteq A_c$ of actions which can be executed in cycle c such that

1. Latency of the design is minimized

$$\text{maximize } |A_c^P|.$$

2. Peak power constraint of the design is met

$$\left(\sum_{i=1 \text{ to } |A_c^P|} p_i \right) \leq P \text{ s.t. } a_i \in A_c^P.$$

3. If $a_i <_d a_j$ holds, then a_i should be given priority over a_j if both a_i and a_j cannot be scheduled in the clock cycle c due to peak power constraint

$$\forall \, a_i, a_j \in T \text{ s.t. } ((i \neq j) \wedge (a_i <_d a_j)),$$
$$a_j \in A_c^P \Rightarrow a_i \in A_c^P.$$

4. If neither $a_i <_d a_j$ nor $a_j <_d a_i$ holds, and $a_i <_s a_j$ holds, then a_i should be given priority over a_j if both a_i and a_j cannot be scheduled in the clock cycle c due to peak power constraint

$$\forall \, a_i, a_j \in T \text{ s.t. } ((i \neq j) \wedge \neg(a_i <_d a_j \vee a_j <_d a_i) \wedge (a_i <_s a_j)),$$
$$a_j \in A_c^P \Rightarrow a_i \in A_c^P.$$

8.3 Peak Power Reduction Algorithm

The proposed peak power reduction algorithm allows the designer to specify a peak power constraint P for a design. A simple version of the algorithm uses the sequential ordering S_{order} (which is based on relation $<_s$) to select a maximal subset $A_c^P \subseteq A_c$ of actions in each clock cycle c under the given peak power constraint. However, implementing a circuitry in hardware in order to dynamically construct A_c^P in each clock cycle c will result in extra power and area overheads. This might undo any peak power reductions obtained by disabling some actions in cycle c. In order to avoid this, the proposed algorithm statically creates a set of all the possible combinations of non-conflicting actions in A which can violate the peak power constraint of the design. Then, for each violating combination A_c, its corresponding non-violating combination A_c^P is determined at compile time. This is done by passing set A as input to Algorithm *CombPairs*.

Algorithm *CombPairs*: Input: *Set R*;   Output: *Set NC*.

Step I. Create a set C of all the possible combinations of non-conflicting actions in R which when executed concurrently will lead to peak power violation in the hardware.

$$C = \{vc_k : vc_k \text{ is a set of non-conflicting actions}$$

$$\text{such that } (\sum_{i=1 \text{ to } |vc_k|} p_i) > P, \ a_i \in vc_k \}$$

Step II. For each peak power violating combination $vc_k \in C$, compute a corresponding non-violating combination nvc_k based on S_{order}. While constructing nvc_k, an action $a_i \in vc_k$ is give preference over any other action $a_j \in vc_k$ occurring later in S_{order}. Return set NC given as,

$$NC = \{(vc_k, nvc_k) : nvc_k \text{ is a non-violating combination}$$

$$\text{corresponding to } vc_k \in C. \}$$

In order to do this analysis, the algorithm first assigns an approximate power-weight p_i to each action $a_i \in A$ of the design. p_i is computed statically based on the number and type of operations occurring in a_i. Larger is the number of operations in an action, larger is its power-weight and hence higher will be its power consumption when executed.

Using set NC returned from Algorithm *CombPairs*, a lookup-table is synthesized in hardware, which when given a violating combination $vc_k \in C$ as input returns the corresponding non-violating combination nvc_k as output. Thus, during the execution of the design, in each clock cycle c set A_c of maximal non-conflicting actions is selected and is passed to this lookup-table as input. The lookup-table checks if the given input matches with any of the stored entries. If yes, then such a combination will violate the peak power constraint of the design, and corresponding non-violating combination A_c^P stored in the table is returned as output, which is then used to execute the appropriate actions satisfying the peak power constraint.

8.3.1 Handling Combinational Path Dependencies

Note that if $a_i <_d a_j$ holds and the peak power algorithm disables a_i in a clock cycle for peak power savings, then the guard of a_j is evaluated again (because of the combinational path) to check if a_j is still enabled in that clock cycle. This will result in a combinational cycle in the hardware because the output of a_j's guard (along with the guards of all the other enabled non-conflicting actions) was already used by the peak power algorithm to select an appropriate set of actions for execution. Such a cycle in hardware can be avoided by refining the peak power reduction algorithm as follows:

1. Use Algorithm *ActionGroups* to group various actions in A such that a_i and a_j lie in separate sets *Set A* and *Set B*. Algorithm *ActionGroups* creates groups of actions based on relation $<_d$. In case both $<_s$ and $<_d$ hold for a_i and a_j in a conflicting manner ($a_i <_d a_j$ and $a_j <_s a_i$), then a_i is selected over a_j because $<_d$ is given priority over $<_s$ during the selection of actions for execution. Depending on the design, multiple such sets for each different group are created by Algorithm *ActionGroups*.

Algorithm *ActionGroups*: Input: $Set\ S_{order}$; **Output:** $Set\ G$.

Step I. $Z = NULL$; $G = NULL$; $S = S_{order}$;
Step II.
if ($S \neq$ NULL) **then**
    Select the first element s of S;
    if ($Z \neq$ NULL) **then**
        if $\exists z_i \in Z : z_i <_d s$ **then**
            $G = G \cup \{Z\}$; $Z = NULL$;
            Go to Step II;
        else if $\exists z_i \in Z : s <_d z_i$ **then**
            $Z = Z - \{z_i\}$; $S = \{z_i\} \cup S$;
            $Z = Z \cup \{s\}$; $S = S - \{s\}$;
            $G = G \cup \{Z\}$; $Z = NULL$;
            Go to Step II;
        end if
    end if
    $Z = Z \cup \{s\}$; $S = S - \{s\}$;
    Go to Step II;
else
    $G = G \cup \{Z\}$;
    return G;
end if

---

2. Use Algorithm *CombPairs* to generate a lookup-table *LUT A* for *Set A* which determines what actions of *Set A* can be allowed to execute under the given peak power constraint P. The inputs to *LUT A* consist of guards of all the actions in *Set A*.

3. Create *Set B$'$* = *Set A* $\cup$ *Set B* and use Algorithm *CombPairs* to generate a lookup-table *LUT B* for *Set B$'$* which determines what actions of *Set B* should be

allowed to execute under the peak power constraint. Inputs to *LUT B* consist of all the outputs of *LUT A* in addition to the guards of all actions in *Set B*.

4. If more than two sets are created by Algorithm *ActionGroups*, then the algorithm continues similarly until lookup-tables for all the sets are generated. For example, if there exists a third set *Set C*, then *Set C′ = Set B′ ∪ Set C* is created and its lookup-table *LUT C* is also generated.

Thus, instead of considering all the non-conflicting actions at once (and deciding which of those should be disabled), the peak power algorithm is refined to first use Algorithm *ActionGroups* to group various actions of a design based on relations $<_d$ and $<_s$. In hardware, using the generated lookup-tables for each group, appropriate actions are disabled for peak power reduction without causing any combinational cycles.

For the Vending Machine design, following groups are created using Algorithm *ActionGroups*:

Set A = { *tenCentIn, fiftyCentIn, doDispenseMoney* }

Set B = { *doDispenseGum, moneyBackButton* }

Note that actions *dispenseTenCents* and *dispenseGum* are not placed in any groups. This is because both these actions do not perform any computations and hence are not important for the purposes of peak power reduction. The proposed algorithm is flexible in the sense that the designer can specify what kind of actions should be excluded from such a low peak power analysis. Such actions will never be disabled for peak power reduction.

8.4 Experiments and Results

We implemented the proposed algorithm in *Bluespec Compiler version 2007.05* (BSC) and tested it on various realistic designs. Each design is first synthesized to RTL using BSC which converts a CAOS-based description of a design to RTL Verilog code. *Synopsys' Design Compiler (version Y-2006.06-SP2 for Linux)*, which is a logic synthesis tool is then used to convert the RTL description of the design to gate-level netlist using the 0.18 μm TSMC technology library. The generated RTL and gate-level Verilog codes are simulated using *Synopsys VCSi (version X-2005.06-9)* to verify the functional behavior of the designs. The synthesized designs are checked to make sure that they meet the timing requirements. Peak power estimation is done both at RTL and at gate level for which the generated Verilog design files and the simulation activity files are passed to *Sequence PowerTheater (version R2006.2)* [99], which is a power analysis tool commonly used in industry.

8.4.1 Designs

The selected designs vary in their nature, size, and complexity. Crossbar Switch is an implementation of a bus/interconnect for a SoC which can be used to connect multiple initiators (like processors, DMA engines) and multiple targets (like memories, I/O blocks). The arbitration between the ports is handled using various actions. IFFT design performs 64-point inverse Fast Fourier Transform on complex frequencies to translate them into the time domain, where they can be transmitted wirelessly (based on 802.11a IEEE standard). LPM (Longest Prefix Match) is a design used in an Internet router to compute the output port to which an input packet should be forwarded based on the destination IP address in the packet header.

8.4.2 Gate-Level Average Power and Peak Power Comparisons

Table 8.1 shows the peak power reductions (as fractional change from original peak power) obtained by applying the proposed algorithm on various designs. In order to evaluate the effect of such peak power reductions on the average power of various designs, Table 8.1 also presents the average power numbers as reported by *PowerTheater*.

Table 8.1 Gate-level power reductions

Design	Original (average power/ peak power) (mW)	Fractional power Reductions using proposed algorithm
Vending machine	36.91/63.30	0.98x/0.93x
Crossbar Switch	97.40/137.80	1.00x/0.78x
IFFT	270.23/496.79	1.02x/0.80x
LPM	123.12/144.93	0.96x/0.91x

Peak power reductions of around 20% in designs like *Crossbar Switch* and *IFFT* demonstrate that the proposed algorithm can be successfully used to reduce the peak power of designs generated using atomic actions. As shown in Table 8.1, average power of various designs varies within small margins. This demonstrates the fact that adding of extra lookup-tables, as part of the proposed algorithm, does not drastically increase the power consumption of a design. This is because under reasonable peak power constraints, generated lookup-tables for most designs are very sparse, and consequently their insertion does not cause intolerable power or area overheads.

8.4.3 Effects on Latency, Area, and Energy

Table 8.2 shows relative changes in the latency (number of clock cycles), area, and energy of various designs on using the proposed algorithm. Latency and area numbers are reported by *Design Compiler*, whereas energy numbers are estimated using the average power and simulation time numbers are reported by *PowerTheater*.

For most designs, change in the latency of the design is within acceptable limits. However, for the IFFT design latency degradation is reported to be 1.7 times. This is because a very strict peak power constraint was specified for the IFFT design. IFFT consists of a three-stage pipeline where the functionality of each stage is described

Table 8.2 Latency, area, and energy overheads of proposed Algorithm

Design	Latency overhead	Area overhead	Energy overhead
Vending machine	1.08x	1.09x	1.05x
Crossbar Switch	1.20x	1.02x	1.20x
IFFT	1.70x	0.98x	1.73x
LPM	1.24x	1.03x	1.19x

using an action. For minimal latency (no peak power constraint), all three actions are allowed to execute concurrently. However, when a strict peak power constraint is specified only one action is allowed to execute (when all three are enabled) by the proposed algorithm. This resulted in an overall increase in its latency. In another experiment, peak power constraint of the IFFT design was relaxed such that two actions can be executed concurrently. In that case, the latency of the design improved at the cost of increase in its peak power. Thus, there is a latency and peak power trade-off associated with the proposed algorithm. Stricter the peak power constraint, larger is the degradation of the latency of a design because less number of actions are allowed to execute concurrently.

Table 8.2 also shows the area overheads associated with the use of proposed algorithm for peak power savings. These area overheads are due to extra lookup-tables generated as part of the algorithm. IFFT actually shows a small decrease in the area of the design which can be attributed to the fact that at times addition of such lookup-tables enables certain area optimizations during the logic synthesis process.

Energy consumed by a hardware design is the product of its average power consumption and the time for which the design is executed. Table 8.2 shows the energy overheads for each design. The latencies have increased for all designs, but the average power numbers did not change much (as shown in Table 8.1), hence the energy numbers show corresponding increase. Thus, based on the experiments, we can conclude that peak power minimization and energy savings are contradictory goals for most hardware designs.

8.4.4 RTL Activity Reduction

PowerTheater also supports the measurement of peak switching activity of a design at RTL. Switching activity of a design is proportional to its power consumption. Table 8.3 shows the reductions obtained in peak switching activity of various designs on using the proposed algorithm.

Table 8.3 Peak switching activity reductions at RTL

Design	Original peak RTL activity (MHz)	Reduced peak RTL activity (MHz)	Fractional change
Vending machine	6.57	5.88	0.89x
Crossbar Switch	2.27	1.59	0.70x
IFFT	3.48	2.62	0.75x
LPM	5.88	5.73	0.97x

Gate-level results are usually considered to be more accurate that RTL results. However, comparing the gate-level peak power reductions reported in Table 8.1 with the peak activity reductions shown in Table 8.3, it can be noticed that for most designs fractional reductions at gate level are similar to those at RTL. Relative latency numbers can also be obtained during RTL simulations. Thus, the proposed algorithm can be used to measure the effects of various peak power constraints on the power and latency of a design earlier in the design cycle (at RTL instead of gate level). This aids in faster architectural exploration by raising the abstraction level at which such comparisons can be done, thus resulting in reduced design time.

8.5 Summary

This chapter presents an algorithm for peak power reduction of designs generated using atomic actions. The proposed algorithm exploits the fact that for designs described using CAOS, disabling appropriate actions in a clock cycle does not alter their functional behavior. Thus, as confirmed by the experimental results, higher abstraction level of such a model of computation inherently offers the advantage of easier power management in hardware designs.

8.6 Issues Related to Proposed Algorithm

1. The proposed algorithm increases the latency of the designs and thus can only be used to reduce the peak power of latency-insensitive designs.
2. For designs with large number of actions, size of the generated lookup-tables can become big if the peak power constraint is too strict. This may cause extra power and area overheads in such designs. If the lookup-tables become bigger and denser, then the use of associative memory might provide a better alternative from area and power point of view.
3. Use of lookup-tables may affect the critical path of the design leading to timing issues in some designs.

Some of the above-mentioned issues can be resolved by refining the proposed algorithm.

Chapter 9
Verifying Peak Power Optimizations Using SPIN Model Checker

Chapter 8 presents an algorithm for peak power reduction of designs generated using CAOS. The proposed algorithm exploits the fact that for a CAOS-based design, disabling appropriate actions in a clock cycle for reducing its peak power should not alter the functional behavior of the design. This is because for well-written CAOS designs, re-scheduling of the actions of a design for reducing its peak power will enable appropriate set of actions for execution (in future clock cycles) based on various fairness constraints, thus maintaining the overall behavior of the design. However, re-scheduling of the actions can result in changing the behavior of those CAOS designs which do not model the fairness constraints (related to the execution of the actions of a design) appropriately. Such designs may fail to produce a correct output when re-scheduling of their actions occurs. This stresses the need for verification of a CAOS-based design in order to ensure that its behavior is maintained after re-scheduling of its actions as done by the peak power reduction algorithm described in Chapter 8. Note that such re-scheduling of the actions of a design can also be done for other reasons such as constraints on the number of resources available.

Other power minimization techniques (for dynamic power as well as peak power reduction) proposed earlier in this book also involve changing the structure or behavior of a CAOS-based design by inserting some additional hardware into the circuit or re-scheduling of the actions of the design. In this chapter, we propose techniques that can be used for the verification of such power-optimized CAOS designs. This work involved investigating various formal verification problems relevant to CAOS-based synthesis and eventually paved the way for a thorough study of these problems and their solutions at a level of abstraction above RTL. The verification techniques presented in this chapter are generic in the sense that they can be used for verifying any kind of changes in the structure or behavior of a CAOS-based design, in addition to the ones caused by various power-minimization techniques that provided the initial motivation for this work.

We consider the following formal verification problems related to CAOS-based designs: (i) Given a CAOS-based specification $\mathscr{S}$ of a hardware design, does it satisfy certain temporal properties? (ii) Given a CAOS-based specification $\mathscr{S}$, and an implementation R synthesized from $\mathscr{S}$ using a CAOS-based synthesis tool, does R conform to the behaviors specified by $\mathscr{S}$; that is, is R a refinement of $\mathscr{S}$?

G. Singh, S.K. Shukla, *Low Power Hardware Synthesis from Concurrent Action-Oriented Specifications*, DOI 10.1007/978-1-4419-6481-6_9,
© Springer Science+Business Media, LLC 2010

(iii) Given a different implementation R' synthesized from $\mathscr{S}$ using some other CAOS-based synthesis tool (for example, an implementation generated using a power-minimization technique), is R' a refinement of R as well? In this chapter, we show how SPIN Model Checker [64] can be used to solve these three problems related to the verification of CAOS-based designs. Using a sample design, we illustrate how our proposed approach can be used for verifying whether the designer's intent in the CAOS-based specification is accurately matched by its synthesized hardware implementation. Thus, techniques proposed in this chapter can be used for the verification of the desired properties of a given CAOS-based design as well as for comparing the behaviors of its various implementations differing in their scheduling of actions.

With regard to verification, it can be argued that various techniques proposed in the past to formally verify RTL implementations of hardware designs can also be used to verify hardware generated from CAOS-based synthesis. However, as more and more functionality is being added to hardware designs, RTL is fast becoming too low level to efficiently handle the complexity of hardware designs, thus making it imperative to investigate techniques for the verification of high-level hardware descriptions earlier in the design cycle. High-level specifications such as CAOS do not contain details irrelevant to the design's behavior, and hence formal verification of such specifications can aid in faster verification and architectural exploration leading to increase in the designer's productivity. Thus, we investigate various verification problems related to CAOS-based designs and propose techniques to solve those problems (at a level of abstraction above RTL) using the SPIN Model Checking tool [64].

9.1 Related Background

SPIN [64] is a model checking tool used for formal verification of distributed software systems. The emphasis in PROMELA, which is the input specification language of SPIN, is on the modeling of process synchronization and coordination [64]. For this reason, SPIN is mainly targeted for the verification of concurrent software systems (described in terms of interacting processes) [63], rather than the verification of hardware designs. On the other hand, design of concurrent hardware systems maps closely to clock-synchronous concurrency model as is primarily implemented by RTL HDLs (Hardware Description Languages) such as Verilog and VHDL. For describing such systems, synchronous state machines are better suited, making the use of tools like SMV [112] more prevalent for hardware verification. However, hardware design flows are changing in nature as demonstrated by high-level synthesis, and due to this changing trend, there is a need for new hardware verification techniques that are suitable for high-level design flows. As mentioned earlier, in this chapter we present techniques which show how SPIN, which is primarily a software verification tool, can be efficiently used for the verification of hardware designs generated from CAOS-based high-level synthesis.

For a given CAOS-based specification $\mathscr{S}$ of a design, its RTL implementation schedules its actions for execution in different clock cycles. A simple hardware schedule randomly selects just one action for execution in each clock cycle. Such a *sequential execution semantics* contains all possible hardware behaviors corresponding to the specification $\mathscr{S}$ but is undesirable from latency (total number of clock cycles elapsed until the execution halts) point of view. Thus, in RTL implementations generated using $\mathscr{S}$, multiple actions are scheduled for execution in a single clock cycle provided the set of behaviors shown by any such implementation is a subset of the set of behaviors of specification $\mathscr{S}$. In this sense, checking the schedule generated by such implementations against the behaviors of the specification $\mathscr{S}$ and figuring out relationships among the behaviors shown by two different implementations of a design are important verification issues.

A hardware design can be represented using a corresponding automaton $\mathscr{A}$ which encodes all the behaviors of the design in terms of its different states and transitions among those states. The language of automaton $\mathscr{A}$ contains all such behaviors of the design and can be denoted as $\mathscr{L}(\mathscr{A})$. Also, essential properties of the behaviors of the design can be expressed as a set of LTL (Linear Temporal Logic) formulae EP written in terms of its states and transitions. Using these notations, important verification problems associated with CAOS-based design flow can be defined as follows:

1. Given a CAOS-based specification $\mathscr{S}$, which corresponds to an automaton $\mathscr{A}_{\mathscr{S}}$ based on the *sequential execution semantics*, does $\mathscr{A}_{\mathscr{S}}$ satisfy all the essential properties of *EP*?
2. Given a CAOS-based specification $\mathscr{S}$ and its implementation R (synthesized using a CAOS-based high-level synthesis tool), which corresponds to an automaton $\mathscr{A}_{\mathscr{R}}$, does $\mathscr{A}_{\mathscr{R}}$ conform to the behaviors of $\mathscr{S}$; that is, does $\mathscr{L}(\mathscr{A}_{\mathscr{R}}) \subseteq \mathscr{L}(\mathscr{A}_{\mathscr{S}})$ hold? In other words, is R a refinement of $\mathscr{S}$?
3. Given automata $\mathscr{A}_{\mathscr{R}}$ and $\mathscr{A}_{\mathscr{R}'}$ for two different implementations R and R' of specification $\mathscr{S}$ (synthesized using two different CAOS-based synthesis tools differing in their scheduling of actions), respectively, is R' a refinement of R; that is, does $\mathscr{L}(\mathscr{A}_{\mathscr{R}'}) \subseteq \mathscr{L}(\mathscr{A}_{\mathscr{R}})$ hold?

In order to solve these three problems efficiently, there is a need for performing automatic formal verification of CAOS-based designs at a level of abstraction above RTL as done by the techniques proposed in this chapter. The two major contributions of this work are as follows: (1) We present formal algorithms for converting a given CAOS-based description of a hardware design and its implementations into corresponding process-based PROMELA models for verification of their essential properties using SPIN. (2) More importantly, we also present a technique which uses SPIN for proving strong language-containment results between two different models of a CAOS-based design. Note that SPIN does not directly support proofs of language-containment between different but related PROMELA models. In this work, we present a technique which generates an LTL specification in the style of TLA (Temporal Logic of Actions) [79] for a given PROMELA model. Such an LTL specification can then be used for proving stronger language-containment results with respect to other related PROMELA models using SPIN's LTL Property Manager.

Thus, we provide a complete automation flow for solving all three verification problems related to CAOS-based hardware design flow. As confirmed by experiments, the proposed approach can be successfully used for quick and early verification of hardware designs generated using CAOS-based synthesis. Throughout the chapter, we use the example of a *Vending Machine* design to illustrate the usefulness of our approach.

The chapter is organized as follows. Section 9.2 presents a formal description of CAOS-based high-level synthesis and various scheduling semantics for CAOS designs. Correctness requirements for CAOS designs are explained in Section 9.3. Algorithms for generating PROMELA models containing scheduling information corresponding to a CAOS-based specification and its implementations are presented with sample models in Section 9.4. Section 9.5 presents algorithms for verification of essential properties of a CAOS design and performing language-containment proofs between CAOS models and also discusses some sample experiments. Section 9.6 concludes the chapter with a short discussion. Finally, we have added an extensive Appendix containing various algorithms and listings of code.

9.2 Formal Description of CAOS-Based High-Level Synthesis

In Chapter 3, we described how hardware can be synthesized from CAOS-based description of a design. In this section, we again present a formal description of some of the important concepts related to CAOS-based high-level synthesis. Notations used in this section will be used throughout this chapter to explain various verification techniques suitable for CAOS-based synthesis.

9.2.1 Hardware Description

Definition 9.1 A CAOS *specification* $\mathscr{S}$ of a hardware design consists of a set $S = \{s_1, s_2, \ldots, s_k\}$ of k state elements of the design and a set $A = \{a_1, a_2, \ldots, a_n\}$ of n actions of the design.

Definition 9.2 Each *state element* $s_i \in S$ is of the form (t_i, n_i, in_i), where t_i, n_i, and in_i represent the data type, name, and initial value of state element s_i, respectively. The state elements of a CAOS design can be in the form of registers, FIFOs, or memories.

Definition 9.3 Each *action* $a_i \in A$ of specification $\mathscr{S}$ consists of two parts: a guard and a body. For example, an action a_i can be of the form

action $m_i() \; if \; g_i(S_g^i)\{s_{i1} \; <= \; b_{i1}(S_b^i); \; s_{i2} \; <= \; b_{i2}(S_b^i); \; s_{i3} \; <= \; b_{i3}(S_b^i); \; \}$ end;

Here, m_i is the name of action $a_i \in A$.

$$S_g^i = \{s : \text{value of } s \in S \text{ is accessed in the guard of } a_i \in A\}.$$
$$S_b^i = \{s : \text{value of } s \in S \text{ is accessed in the body of } a_i \in A\}.$$

Definition 9.4 g_i denotes the *guard* of action a_i. It is a condition associated with a_i which evaluates to either *True* or *False* depending on current values of elements of S_g^i. Action a_i is said to be *enabled* if g_i evaluates to *True*.

Definition 9.5 *Body* of action a_i consists of set of assignment statements of the form $s_{ij} <= b_{ij}(S_b^i)$, computing next state values of the design and, in general, it can be expressed as

$$B_i = \{(s_{ij}, b_{ij}(S_b^i)) : b_{ij}(S_b^i) \text{ computes the next state value for } s_{ij} \in S_u^i$$
$$\text{using current values of the elements of } S_b^i\}$$

where $S_u^i = \{s : \text{value of } s \in S \text{ is updated in the body of } a_i \in A\}$.

Thus, B_i denotes the body of action $a_i \in A$. B_i consists of a group of operations all of which are executed atomically if a_i is enabled and selected for execution.

For better understanding of CAOS semantics and associated verification issues, in this chapter we will explain our verification approach with respect to the *Vending Machine (VM)* design shown in Fig. 8.1. Let $A_m \subseteq A$ be the set of actions corresponding to the interface methods of the CAOS design. Thus, for the VM specification of Fig. 8.1, we have

$S = \{count, moneyBack\}.$

$A = \{doDispenseMoney, doDispenseGum, tenCentIn, fiftyCentIn,$

$\qquad moneyBackButton, dispenseTenCents, dispenseGum\}.$

$A_m = \{tenCentIn, fiftyCentIn, moneyBackButton, dispenseTenCents, dispenseGum\}.$

9.2.2 Scheduling of Actions

Let $d(s_i)$ denote the domain of state element $s_i \in S$ of the design. For hardware designs, usually $d(s_i)$ is the Boolean domain but for CAOS-based specification, which is at a higher level of abstraction, domains such as 32-bit integer, n-bit bit vector can be used. For example, if s_i is an 8-bit register, then $d(s_i) = \{0, 1\}^8$, which is a set of all possible 8-bit strings of 0s and 1s.

Let us consider a vector $\hat{s} =< s_1, s_2, \ldots, s_k >$ of the state elements of the design. Note that $\hat{s}$ contains same elements as in set S. In this section, instead of S, we will use $\hat{s}$ in order to appropriately denote the state of the design for defining its behaviors. Let $\sigma_c(\hat{s}) =< d_1, d_2, \ldots, d_k > \in \prod_{i=1}^{k} d(s_i)$ denote the state of the design at the end of clock cycle c. Thus, σ is the function which maps the state elements of the design to their respective values at some point during the design execution. A behavior of the design can be defined as a sequence of states (possibly infinite) given as $\beta = (\sigma_0(\hat{s}), \sigma_1(\hat{s}), \sigma_2(\hat{s}), \ldots, \sigma_c(\hat{s}), \ldots)$ such that $\sigma_{c+1}(\hat{s})$ results from executing actions in $A_c \subseteq A$ in clock cycle $(c+1)$ when the design is in state $\sigma_c(\hat{s})$.

Refinement of Behaviors – Consider two behaviors β and β' of the design. Let C_β and $C_{\beta'}$ denote the set of clock cycles of β and β', respectively, such that $C_\beta \subset \mathcal{N}$ and $C_{\beta'} \subset \mathcal{N}$, where $\mathcal{N}$ denotes the set of natural numbers. β' is said to be a refinement of β if $\exists\, r : C_\beta \rightarrow C_{\beta'}$, where r is an injective and monotonic function such that $\forall\, c \in C_\beta$, $\sigma_c^\beta(\hat{s}) = \sigma_{r(c)}^{\beta'}(\hat{s})$.

9.2.2.1 AOA Semantics

During the execution of the design generated using CAOS-based synthesis, multiple actions can get enabled in a clock cycle c. In a simple hardware schedule, one such enabled action can be randomly chosen for execution in c, thus proceeding the execution of the design in a sequential manner. The execution of the design halts when none of its actions are enabled in some clock cycle. We call such a sequential execution semantics where only one action is randomly chosen for execution in each clock cycle as *Any One Action (AOA) Semantics*.

Behavior in AOA Semantics – Behavior of the design in *AOA Semantics* can be given as $\beta^{AOA} = (\sigma_0(\hat{s}), \sigma_1(\hat{s}), \sigma_2(\hat{s}), \ldots, \sigma_c(\hat{s}), \ldots)$ such that $\sigma_{c+1}(\hat{s})$ results from executing actions in $A_c \subseteq A$, $|A_c| = 1$, in clock cycle $(c+1)$ when the design is in state $\sigma_c(\hat{s})$.

9.2.2.2 Concurrent Semantics

In spite of being behaviorally correct, the sequential execution of just one action in each clock cycle as per *AOA Semantics* is undesirable from a latency point of view, especially for designs containing large number of actions. Thus, in a hardware implementation R, synthesized from specification $\mathscr{S}$, multiple enabled sets of actions $A_c \subseteq A$, $|A_c| \geq 1$, can be allowed to execute concurrently in a clock cycle c provided the atomicity of actions in A_c is maintained. This means that, in implementation R, behavior of the design resulting from concurrent execution of actions in A_c should be equivalent to at least one sequential behavior of actions in A_c based on *AOA Semantics*. We call such a scheduling semantics where multiple actions are allowed to execute concurrently in a single hardware clock cycle as *Concurrent Semantics*.

Conflicting Actions – In hardware generated from CAOS-based synthesis, maintaining such atomicity among various actions belonging to A_c may lead to complicated combinational circuit. To avoid this, a notion of *conflict* is introduced. An example of a conflict is two actions updating the same hardware register; that is, two actions $a_i, a_j \in A$ can be said to be conflicting with each other if $S_u^i \cap S_u^j \neq \phi$.

Other kinds of conflicts can also exist within two actions of the design, thus forbidding the concurrent execution of those actions. In general, two actions are considered to be conflicting with each other if executing their operations in the same clock cycle is undesirable for pragmatic reasons (like long critical paths, write–write conflicts, complicated hardware analysis). In the synthesized circuit, such restrictions are enforced using small overhead logic.

For example, for the VM design (Fig. 8.1), actions *doDispenseGum*, *tenCentIn*, and *fiftyCentIn* conflict with each other since they all update register *count*. In case two or more conflicting actions are enabled in the same clock cycle, a notion of *priority* is used to decide which of those actions should be executed in that cycle. A higher priority action is always chosen for execution over all the other lower priority conflicting actions. Let $C(i, j)$ represent a function which returns *True* if two actions $a_i, a_j \in A$ conflict with each other, and *False* otherwise. Then, $C_i = \{ a_j : C(i, j) = True; a_j \in A$ has higher priority than $a_i \in A \}$ denotes the set of actions conflicting with a_i which are preferred for execution over a_i.

Sequential Ordering – In order to generate appropriate scheduling and control logic that maintains the atomicity of various actions of a design executing within the same clock cycle, CAOS-based synthesis involves constructing (at compile time) a single sequential ordering S_{order} of all actions belonging to A of specification $\mathscr{S}$. Let us define a relation $<_s$ among any two actions $a_i, a_j \in A$, $C(i, j) = False$, such that $a_i <_s a_j$ holds if concurrent execution of a_i and a_j in a single clock cycle is equivalent to executing a_i followed by a_j in two consecutive clock cycles each executing just one action. For the VM design (Fig. 8.1), actions *doDispenseGum* and *moneyBackButton* can be executed concurrently since their concurrent execution is equivalent to the following sequential ordering: *doDispenseGum, moneyBackButton*. This can be denoted as *doDispenseGum* $<_s$ *moneyBackButton*.

To construct S_{order}, transitivity property of relation $<_s$ is used and cycles are broken appropriately during the synthesis process. For the VM design (Fig. 8.1), one such possible ordering S_{order} of actions is given as *tenCentIn, fiftyCentIn, doDispenseMoney, doDispenseGum, moneyBackButton*. Note that for simplification, we ignore actions *dispenseTenCents* and *dispenseGum* in this ordering since these actions neither perform any computations nor change the state of the design.

Behavior in Concurrent Semantics – For an implementation R, which is synthesized from specification $\mathscr{S}$ based on *Concurrent Semantics*, a behavior of the design can be defined as $\beta^R = (\sigma_0(\hat{s}), \sigma_1(\hat{s}), \sigma_2(\hat{s}), \ldots, \sigma_c(\hat{s}), \ldots)$, such that

(1) $\sigma_{c+1}(\hat{s})$ results from executing actions belonging to $A_c \subseteq A$ in clock cycle $(c + 1)$ when the design is in state $\sigma_c(\hat{s})$.
(2) If $|A_c| > 1$, then $C(i, j) = False \forall a_i, a_j \in A_c, i \neq j$; that is, A_c denotes a set of non-conflicting actions.
(3) β^R corresponds to an equivalent behavior β^R_{seq} generated using the sequential ordering S_{order} of R, with just one action being executed in each clock cycle in β^R_{seq}.

9.3 Correctness Requirements for CAOS Designs

9.3.1 AOA Semantics

Depending on what actions are selected for execution in different clock cycles, specification $\mathscr{S}$ of the design can consist of multiple behaviors of the form β^{AOA} based on *AOA Semantics*. Let $\mathscr{A}_{\mathscr{S}}$ be the automaton encoding all such possible behaviors of specification $\mathscr{S}$. The language of automaton $\mathscr{A}_{\mathscr{S}}$ is denoted by $\mathscr{L}(\mathscr{A}_{\mathscr{S}})$ and is said to contain all behaviors of specification $\mathscr{S}$. Let EP represent the set of all essential properties of the design expressed as LTL (Linear Temporal Logic) formulae.

   Correctness Requirement 1 (CR-1). The correctness constraint mandates that for $\mathscr{S}$ to be a valid specification of the hardware design, each behavior β^{AOA} of $\mathscr{S}$ should satisfy all properties in EP; that is, $\forall\ \beta^{AOA} \in \mathscr{L}(\mathscr{A}_{\mathscr{S}}), \forall\ p \in EP, \beta^{AOA}$ should satisfy p.

9.3.2 Concurrent Semantics

For an implementation R generated from specification $\mathscr{S}$ based on *Concurrent Semantics*, any behavior of the form β^{R} shown by R corresponds to an equivalent behavior β^{R}_{seq} generated using S_{order}. Let A_{R} be an automaton encoding all possible behaviors of the form β^{R}_{seq} shown by R.

   Correctness Requirement 2 (CR-2). The correctness constraint mandates that for R to be a valid implementation of $\mathscr{S}$, R should be a refinement of S. In other words, language-containment relation $\mathscr{L}(\mathscr{A}_{\mathscr{R}}) \subseteq \mathscr{L}(\mathscr{A}_{\mathscr{S}})$ should hold.

9.3.2.1 Maximal Concurrent Schedule (MCS)

For latency minimization, maximal set of actions of the design can be chosen for execution in each hardware clock cycle. Such a schedule of a design can be termed as a *Maximal Concurrent Schedule (MCS)* and an implementation R_{MCS} of the design generated based on this schedule contains multiple different behaviors of the form β^{R} such that $A_{c} = A_{c}^{M}$, where $A_{c}^{M} \subseteq A$ is a maximal set of non-conflicting actions scheduled for execution in clock cycle c. As mentioned earlier, each behavior β^{R} corresponds to an equivalent behavior β^{R}_{seq} generated using the sequential ordering S_{order} of R_{MCS}.

   Let $\mathscr{A}_{\mathscr{R}}^{\mathscr{MCS}}$ be the automaton encoding all such possible behaviors $\mathscr{L}(\mathscr{A}_{\mathscr{R}}^{\mathscr{MCS}})$ of a hardware design under the maximal concurrent schedule. The correctness constraint requires that the language-containment relation $\mathscr{L}(\mathscr{A}_{\mathscr{R}}^{\mathscr{MCS}}) \subseteq \mathscr{L}(\mathscr{A}_{\mathscr{S}})$ should hold; that is, R_{MCS} should be a refinement of specification $\mathscr{S}$ of the design. Bluespec Compiler (BSC) performs automatic concurrent scheduling of hardware designs and generates RTL code adhering to one such maximal concurrent refinement which satisfies $\mathscr{L}(\mathscr{A}_{\mathscr{R}}^{\mathscr{MCS}}) \subseteq \mathscr{L}(\mathscr{A}_{\mathscr{S}})$. (Bluespec System Verilog (BSV) is the CAOS-style input specification language of BSC.)

9.3.2.2 Alternative Concurrent Schedule (ACS)

As mentioned earlier, BSC schedules maximal set of actions in each clock cycle for latency minimization. However, concurrent execution of large number of actions for improving the latency of a hardware design is usually associated with a corresponding degradation of other attributes of the design, such as its area, peak power. This might not be desirable for a design having conflicting constraints on its latency and other attributes.

In such cases, instead of executing maximal set of actions A_c^M in clock cycle c, an alternative implementation R_{ACS} of the design needs to be derived which selects only a set of actions $A_c \subseteq A$ for execution in c such that all the constraints of the design are satisfied. Such a schedule of a design can be termed as a *Alternative Concurrent Schedule (ACS)*. Let $\mathscr{A}_{\mathscr{R}}^{\mathscr{ACS}}$ be the automaton encoding all possible behaviors $\mathscr{L}(\mathscr{A}_{\mathscr{R}}^{\mathscr{ACS}})$ of a hardware design corresponding to R_{ACS}. Again, the correctness constraint mandates that $\mathscr{L}(\mathscr{A}_{\mathscr{R}}^{\mathscr{ACS}}) \subseteq \mathscr{L}(\mathscr{A}_{\mathscr{S}})$; that is, any alternative implementation R_{ACS} of a design based on such a schedule is required to be a refinement of specification $\mathscr{S}$ of the design.

9.3.3 Comparing Two Implementations

Two different implementations of a CAOS-based design differ in their scheduling of the actions of the design. In general, an implementation R of a CAOS-based specification $\mathscr{S}$ may either enhance or restrict its set of behaviors as compared to some other implementation R' (as an example, R can be generated using a latency-optimized synthesis tool whereas R' can be generated using a power-optimized one). However, as mentioned earlier, all behaviors of R and R' should conform to the set of behaviors $\mathscr{L}(\mathscr{A}_{\mathscr{S}})$ of the specification $\mathscr{S}$, thus satisfying $\mathscr{L}(\mathscr{A}_{\mathscr{R}}) \subseteq \mathscr{L}(\mathscr{A}_{\mathscr{S}})$ and $\mathscr{L}(\mathscr{A}_{\mathscr{R}'}) \subseteq \mathscr{L}(\mathscr{A}_{\mathscr{S}})$.

Furthermore, depending on the design requirements, in some cases it might also be desirable to show that R' is a refinement of R; that is, $\mathscr{L}(\mathscr{A}_{\mathscr{R}'}) \subseteq \mathscr{L}(\mathscr{A}_{\mathscr{R}})$ holds as shown in Fig. 9.1a. For other cases, relation shown in Fig. 9.1b may hold.

Correctness Requirement 3 (CR-3). For two different valid implementations R and R' of specification $\mathscr{S}$, R' is a refinement of R iff $\mathscr{L}(\mathscr{A}_{\mathscr{R}'}) \subseteq \mathscr{L}(\mathscr{A}_{\mathscr{R}})$.

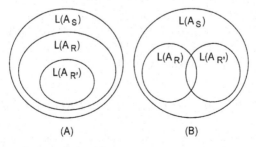

Fig. 9.1 Language-containment relationships

9.4 Converting CAOS Model to PROMELA Model

As mentioned earlier, formal verification of a CAOS-based design is mandatory in order to ensure that its specification and various implementations adhere to designer's intentions of the design behavior and meet all the correctness requirements. Different hardware scheduling semantics related to CAOS-based synthesis can be efficiently modeled in PROMELA using various constructs of the language [64]. Hence, as demonstrated by the techniques presented in this chapter, PROMELA can be used for modeling the desired semantics of CAOS designs, thus providing a path for verification of such designs using SPIN at a level of abstraction above RTL.

9.4.1 Why SPIN?

We propose the use of SPIN for verification of CAOS-based designs because of the following reasons:

1. PROMELA supports the description of concurrent systems at the desired level of abstraction.

   a. Atomicity of actions is an important concept in CAOS. Such atomicity of operations is well supported in PROMELA and can be used to model a CAOS design.
   b. As demonstrated by our approach, different scheduling semantics and priorities among various operations of CAOS-based designs can also be efficiently modeled using PROMELA constructs like "unless," "do" [64].

2. SPIN supports the verification of properties of a PROMELA model. As shown in this work, SPIN's LTL Property Manager can be efficiently used for proving strong language-containment relations between different CAOS models.

9.4.2 Generating PROMELA Variables and Processes

PROMELA model $\mathcal{M}$ of a system consists of a set V of variables and a set P of processes (P includes the "init" process) used to describe the system. Given a CAOS-based specification $\mathcal{S}$ of a hardware design, we present Algorithm *GenPROMELA* (Fig. 9.2) to generate PROMELA model $\mathcal{M}$ corresponding to $\mathcal{S}$. Algorithms *GenVARS, GenPROCS,* and *GenProcCycle* which are used by Algorithm *GenPROMELA* are explained in the Appendix.

9.4.3 Adding Scheduling Information to PROMELA Model

Algorithm *GenPROMELA* (Fig. 9.2) generates sets of variables and processes for PROMELA model $\mathcal{M}$ which corresponds to specification $\mathcal{S}$. For modeling a

---

ALGORITHM: *GenPROMELA*. **INPUT:** CAOS-based Specification $\mathscr{S}$.
OUTPUT: PROMELA model $\mathscr{M}$.

---

1. Initialize $V = \phi$, $P = \phi$. (**Note:** V and P are sets of variables and processes of PROMELA model $\mathscr{M}$ respectively.)
2. Using Algorithm *GenVARS* (Appendix - Figure 9.5), generate the set of variables V for $\mathscr{M}$ using $\mathscr{S}$. Algorithm *GenVARS* generates variables corresponding to the state elements of the CAOS design as well as other variables needed for modeling the concurrent hardware behavior of the design.
3. Using Algorithm *GenPROCS* (Appendix - Figure 9.6), generate the set of processes P for $\mathscr{M}$ using $\mathscr{S}$ and V. For each action of $\mathscr{S}$, Algorithm *GenPROCS* generates a corresponding process in $\mathscr{M}$ modeling the behavior of the action.
4. In order to model the hardware behavior in PROMELA, use Algorithm *GenProcCycle* (Appendix - Figure 9.7) to generate a process pr using $\mathscr{S}$, V and P. Add pr to P. The execution of this process is used to denote the start of a hardware cycle, thus modeling the synchronous execution of a hardware design.
5. Using V and P, generate PROMELA 'init' process which initializes all variables in V and instantiates all processes in P. Add 'init' to P
6. Output PROMELA model $\mathscr{M}$ whose sets of variables and processes are denoted by V and P respectively.

Fig. 9.2 Algorithm for generating PROMELA model from CAOS-based specification

particular hardware execution semantics in PROMELA, a new model $\mathscr{M}_f$ needs to be generated by enhancing model $\mathscr{M}$ with the corresponding hardware scheduling information.

9.4.3.1 AOA Semantics

In order to model a schedule based on *AOA Semantics* in PROMELA, we present Algorithm *AddSeqSched* (Appendix, Fig. 9.8). The algorithm enhances PROMELA model $\mathscr{M}$ generated by Algorithm *GenPROMELA* such that during the execution of the model, "start_of_cycle" process (whose execution denotes the start of a new hardware clock cycle) is executed after every execution of any other process of the model. The generated model $\mathscr{M}_f$ accurately models the behaviors $\mathscr{L}(\mathscr{A}_{\mathscr{S}})$ of specification $\mathscr{S}$ as per *AOA Semantics*.

9.4.3.2 Concurrent Semantics

During the CAOS-based synthesis, a sequential ordering S_{order} of the actions of the design is generated to which any concurrent execution of actions will correspond. Given such an ordering S_{order} for an implementation R, we present an Algorithm *AddConcSched* (Appendix, Fig. 9.9), which generates model $\mathscr{M}_f$ by enhancing PROMELA model $\mathscr{M}$ (generated by Algorithm *GenPROMELA* shown in Fig. 9.2) with the scheduling information of implementation R. Note that behaviors of model $\mathscr{M}_f$ correspond to all possible behaviors $\mathscr{L}(\mathscr{A}_{\mathscr{R}})$ of R.

Algorithm *AddConcSched* is generic in the sense that it can model any particular schedule (MCS as well as ACS) of a design based on *Concurrent Semantics*. For this, it takes ordering S_{order} corresponding to R and maximum number of actions allowed to execute concurrently in R as inputs. In order to model the hardware behavior, after every execution of "start_of_cycle" process, Algorithm *AddConcSched* checks each process for execution based on S_{order} until maximum number of processes have executed.

9.4.4 Sample PROMELA Models

In the Appendix at the end of the chapter, we show generated PROMELA models (Listings 9.1 and 9.2) corresponding to the implementations of VM specification of Fig. 8.1. Listings 9.1 and 9.2 show two PROMELA models – one corresponding to the implementation R_{MCS} which executes maximal set of actions in a single clock cycle and another corresponding to implementation R_{ACS} based on an alternative schedule which executes only one action as per a sequential ordering.

These models are generated using Algorithm *GenPROMELA* (Fig. 9.2) and Algorithm *AddConcSched* (Appendix, Fig. 9.9). Both the PROMELA models are shown in Listings 9.1 and 9.2 using appropriate markings for implementation-specific lines of code. These models consist of multiple processes including the PROMELA "init" process and five other processes corresponding to different actions of the VM specification. Not all processes could fit in Listing 9.1 so the remaining ones are shown in Listing 9.2. The main characteristics of such a conversion process that translates a given CAOS model into corresponding PROMELA model are as follows:

1. Each action of a CAOS design with its corresponding guard and operations is modeled as a process in PROMELA. Moreover, as shown in Listing 9.1, set of variables V (corresponding to the state elements of VM design) are declared in the beginning of the PROMELA code. Variables *count_old, action_fired*, and *one_action_fired* are used for verification purposes.
   In order to model the atomicity of operations of an action, PROMELA construct "atomic" is used [64]. This avoids the interleaving of operations of various processes, which is consistent with CAOS semantics and aids in faster verification. Also, "do," which is a repetition construct in *PROMELA* [64], is used for forwarding the execution of processes as in the real hardware (cycle by cycle).
2. In order to model the hardware behavior, an extra process named "start_of_cycle" is generated as shown in Listing 9.2. This process denotes the start of a hardware clock cycle and reads inputs, if any, from environment external to the design (like a testbench or another hardware design). For the VM design, such external inputs are read in variables *tenCentIn, fifty-CentIn*, and *moneyBackButton*. These variables are used to signal if processes

IFC_tenCentIn, IFC_fiftyCentIn, and *IFC_moneyBackButton* which correspond to interface methods of the VM specification are executed or not.

3. For processes which do not correspond to interface methods, the execution is dependent on a condition which contains logic related to the guard of the corresponding action of the design, as well as logic based on the conflicts with other higher priority actions. If this condition is *True*, then corresponding process is executed, otherwise next process in the ordering is considered for execution (as controlled by variable *action* and PROMELA's "unless" construct in Listings 9.1 and 9.2).

4. For implementations based on *Concurrent Semantics*, variable named *action* is used to enforce a particular sequential ordering S_{order} of execution among the processes of the generated PROMELA model such that its behavior maps exactly to the concurrent hardware behavior. This is implemented using Algorithm *Add-ConcSched* (Appendix, Fig. 9.9).

Maximal Concurrent Schedule (MCS) – As shown in Listings 9.1 and 9.2 of the Appendix, in order to model an implementation based on MCS, each execution of a process in PROMELA code assigns a new value to variable *action*. The new value is assigned such that in the next step, next process in S_{order} is checked for execution. Such assignments are shown with lines of code marked as "FOR MAXIMAL CONCURRENT SCHEDULE" in Listing 9.1. In such a model, all the processes of the model are checked for execution in every clock cycle.

Alternative Concurrent Schedule (ACS) – For the VM design, an implementation corresponding to MCS will execute actions *doDispenseGum* and *moneyBackButton* concurrently whenever *count* becomes greater than 50 cents, *moneyBack* is *False*, and external environment signals the execution of action *moneyBackButton*. However, if the peak-power constraint of the design allows only one action to execute in a single clock cycle, then an alternative implementation of the VM specification adhering to the peak-power constraint needs to be generated. Listings 9.1 and 9.2 also show generated PROMELA model corresponding to such an implementation of the VM design. For that implementation, appropriate assignments to variable *action* are shown with lines of code marked as "FOR ALTERNATIVE CONC SCHEDULE." The value of variable *action* is updated to six in each process, thus signifying the end of a hardware clock cycle after one process executes.

Note: Enforcing a particular sequential ordering (as done in Algorithm *AddConcSched*) suppresses non-determinism in the behavior of the PROMELA model but is needed to model the deterministic hardware behavior. However, note that no such sequential ordering is enforced in the PROMELA model generated by Algorithm *AddSeqSched* (Appendix, Fig. 9.8). In that model, execution of a single process marks the end of a hardware clock cycle, and in the next cycle, a new process is non-deterministically (and not based on a sequential ordering) executed. Thus, such a model will contain all possible behaviors $\mathcal{L}(\mathcal{A}_{\mathcal{S}})$ of a CAOS-based specification $\mathcal{S}$.

9.5 Formal Verification Using SPIN

9.5.1 Verifying Correctness Requirement 1 (CR-1)

Proposition 9.1 *Given a set EP of essential properties, a CAOS-based specification $\mathscr{S}$ satisfies property $p \in EP$ iff its corresponding PROMELA model $\mathscr{M}_f$ satisfies property p_m, where p_m is equivalent to p and is expressed with respect to $\mathscr{M}_f$.*

Based on Proposition 9.1, a CAOS-based specification $\mathscr{S}$ can be verified for *Correctness Requirement 1 (CR-1)* mentioned in Section 9.3 using Algorithm *VerfCR1* (Fig. 9.3).

ALGORITHM: *VerfCR1.*
INPUT: 1. CAOS-based specification $\mathscr{S}$, 2. Set EP of Essential Properties.
OUTPUT: Verify if $\mathscr{S}$ meets *Correctness Requirement 1 (CR-1)?*

---

1. Using Algorithm *GenPROMELA* (Figure 9.2) and Algorithm *AddSeqSched* (Appendix – Figure 9.8), generate a PROMELA model $\mathscr{M}_f$ based on *AOA Semantics* using $\mathscr{S}$.
2. Initialize $EP_m = \phi$.
3. For each property $p \in EP$

   a. Convert p into p_m such that p_m is an LTL formula expressed with respect to $\mathscr{M}_f$ (using variable set V of $\mathscr{M}_f$).
   b. Add p_m to EP_m.

4. $\forall\, p_m \in EP_m$, perform verification of $\mathscr{M}_f$ against p_m using SPIN.
5. If verification is successful $\forall\, p_m \in EP_m$, then $\mathscr{S}$ meets *Correctness Requirement 1*.

Fig. 9.3 Algorithm for verifying correctness requirement 1

9.5.2 Verifying Correctness Requirement 2 (CR-2)

Proposition 9.2 *Given a CAOS-based specification $\mathscr{S}$ and its implementation R, R is a refinement of $\mathscr{S}$ iff $\mathscr{M}_f^{\mathscr{R}}$ is a refinement of $\mathscr{M}_f^{\mathscr{S}}$, where $\mathscr{M}_f^{\mathscr{R}}$ and $\mathscr{M}_f^{\mathscr{S}}$ are corresponding PROMELA models of R and $\mathscr{S}$, respectively.*

Based on Proposition 9.2, an implementation R of a CAOS-based specification $\mathscr{S}$ can be verified for *Correctness Requirement 2* (Section 9.3) using Algorithm *VerfLangCont* (Fig. 9.4). Algorithm *VerfLangCont* generates PROMELA model $\mathscr{M}_f^{\mathscr{R}}$ for implementation R. It also generates an LTL specification LTL_S encoding all the behaviors $\mathscr{L}(\mathscr{A}_{\mathscr{S}})$ of specification $\mathscr{S}$ (based on *AOA Semantics*) with respect to the state in $\mathscr{M}_f^{\mathscr{R}}$. Such LTL specifications are generated in the style of TLA [79] (as an example, see Listing 9.3 which is shown in the Appendix and is explained later in this chapter) because CAOS-based specifications are written in terms of actions and not explicitly in terms of the state of the design. However, note that Algorithm *VerfLangCont* only encodes the safety properties and not the liveness assumptions in the generated LTL specification LTL_S. This is because we are interested in showing that safety properties encoded in LTL_S also hold in implementation R. In other words, R is a refinement of $\mathscr{S}$.

Using SPIN's LTL Property Manager, model $\mathscr{M}_f^{\mathscr{R}}$ is then verified against LTL_S for ensuring the language-containment relation $\mathscr{L}(\mathscr{A}_{\mathscr{R}}) \subseteq \mathscr{L}(\mathscr{A}_{\mathscr{S}})$. This allows using SPIN for proving strong language-containment relationships for CAOS designs.

9.5.3 Verifying Correctness Requirement 3 (CR-3)

Proposition 9.3 *Given two different implementations R and R' of a CAOS-based specification $\mathscr{S}$, R' is a refinement of R iff $\mathscr{M}_f^{\mathscr{R}'}$ is a refinement of $\mathscr{M}_f^{\mathscr{R}}$, where $\mathscr{M}_f^{\mathscr{R}'}$ and $\mathscr{M}_f^{\mathscr{R}}$ are corresponding PROMELA models of R' and R, respectively.*

Based on Proposition 9.3, Algorithm *VerfLangCont* (Fig. 9.4) can also be used to verify *Correctness Requirement 3* (Section 9.3). For this, all the steps of the Algorithm *VerfLangCont* remain same except that the inputs to the algorithm in this case will be implementations R and R' (instead of specification $\mathscr{S}$ and R as shown in

---

ALGORITHM: *VerfLangCont*.
INPUT: 1. CAOS-based specification $\mathscr{S}$, 2. Implementation R of $\mathscr{S}$.
OUTPUT: Verify if $\mathscr{L}(\mathscr{A}_{\mathscr{R}}) \subseteq \mathscr{L}(\mathscr{A}_{\mathscr{S}})$ holds or not?

---

1. Using Algorithm *GenPROMELA* (Figure 9.2), generate PROMELA model $\mathscr{M}^{\mathscr{R}}$ for implementation R. Let V and P denote the set of variables and processes corresponding to $\mathscr{M}^{\mathscr{R}}$ respectively.
2. Initialize LTL_S = True.
3. For each process $pr \in P$, such that $pr \equiv (m, B, g)$ corresponds to an action $a \in A$,

   a. Generate a set NSV of all next state values in $\mathscr{M}^{\mathscr{R}}$ that are possible according to the behaviors of $\mathscr{S}$ (AOA Semantics) when g becomes *True* in $\mathscr{M}^{\mathscr{R}}$. (**Note:** Only next state values in $\mathscr{M}^{\mathscr{R}}$ corresponding to the execution of an action in $\mathscr{S}$ need to be captured. For this, in order to signify the execution of a process in $\mathscr{M}^{\mathscr{R}}$, value of variable *action_fired* in V can be checked to be *True*.)
   b. Generate an LTL property expression LTL_p of the form $LTL_p \equiv ([ ] (g -> X(\| NSV)))$, where ($\| NSV$) is *True* only when at least one of the elements of NSV is *True*, and *False* otherwise.
   c. $LTL_S = LTL_S$ && LTL_p.

4. Optimize LTL_S to reduce the number of different LTL expressions while retaining all its behaviors. (**Note:** At this stage, LTL_S contains all possible behaviors of $\mathscr{S}$ with respect to the state in $\mathscr{M}^{\mathscr{R}}$.)
5. Using Algorithm *AddConcSched* (Appendix – Figure 9.9), generate PROMELA model $\mathscr{M}_f^{\mathscr{R}}$ using $\mathscr{S}$, $\mathscr{M}^{\mathscr{R}}$ and R.
6. Using SPIN's LTL Property Manager, perform verification of $\mathscr{M}_f^{\mathscr{R}}$ against LTL_S. If verification is successful, then $\mathscr{L}(\mathscr{A}_{\mathscr{R}}) \subseteq \mathscr{L}(\mathscr{A}_{\mathscr{S}})$ holds, otherwise not.

Fig. 9.4 Algorithm for proof of language-containment

Fig. 9.4). Consequently, the algorithm will verify PROMELA model of implementation R' against the LTL specification encoding all the behaviors of implementation R with respect to state in R'.

9.5.4 Sample Experiments

I. We used Algorithm *VerfLangCont* (Fig. 9.4) to successfully verify the language-containment relations $\mathscr{L}(\mathscr{A}_{\mathscr{R}}^{\mathscr{MCS}}) \subseteq \mathscr{L}(\mathscr{A}_{\mathscr{S}})$ and $\mathscr{L}(\mathscr{A}_{\mathscr{R}}^{\mathscr{ACS}}) \subseteq \mathscr{L}(\mathscr{A}_{\mathscr{S}})$ among VM specification of Fig. 8.1 and its implementations R_{MCS} and R_{ACS}, shown in Listings 9.1 and 9.2 of the Appendix. Listing 9.3 of the Appendix shows the LTL specification LTL_S generated by Algorithm *VerfLangCont*, which corresponds to all behaviors $\mathscr{L}(\mathscr{A}_{\mathscr{S}})$ of the VM specification. In Listing 9.3, variables *count_old*, *action_fired*, and *one_action_fired* are used for expressing LTL_S with respect to the state in the PROMELA models of Listings 9.1 and 9.2 as follows:

1. *count_old* stores the old value of *count* at the start of every process and is used in LTL_S to compare any updates on the value of *count* (during the execution of the process) with the old value.
2. *action_fired* and *one_action_fired* are used in LTL_S to check the state of the PROMELA model at points (just after a process has executed its atomic block) which map to the state changes during the execution of the CAOS design.

II. We also used Algorithm *VerfLangCont* to verify if $\mathscr{L}(\mathscr{A}_{\mathscr{R}}^{\mathscr{MCS}}) \subseteq \mathscr{L}(\mathscr{A}_{\mathscr{R}}^{\mathscr{ACS}})$ holds for the VM design.

Result – Verification done by SPIN proved that $\mathscr{L}(\mathscr{A}_{\mathscr{R}}^{\mathscr{MCS}}) \subseteq \mathscr{L}(\mathscr{A}_{\mathscr{R}}^{\mathscr{ACS}})$ does not hold for the VM design and pointed out a behavior shown by R_{MCS} which is not shown by R_{ACS}. This behavior corresponds to the case when $count \geq 100$ holds, *moneyBack* is *False*, and external environment signals the execution of action *moneyBackButton* (assuming actions *tenCentIn* and *fiftyCentIn* are not signaled to execute). In such a state, the behaviors of the two implementations R_{MCS} and R_{ACS} differ as follows:

1. R_{MCS} (to which the generated PROMELA model corresponds) executes actions *doDispenseGum* and *moneyBackButton* in the first clock cycle. This is followed by the execution of *doDispenseMoney* in the next clock cycle.
2. R_{ACS} (to which the generated LTL specification corresponds) only executes *doDispenseGum* in the first clock cycle. This is because R_{ACS} allows only one action to execute in a single clock cycle, and based on the sequential ordering it just executes action *doDispenseGum*, thus dispensing gum and reducing *count* by 50 cents. In the next clock cycle, $count \geq 50$ holds and *doDispenseGum* will be again selected for execution (assuming actions *tenCentIn* and *fiftyCentIn* are not signaled to execute). Action *moneyBackButton* is not executed in such a behavior.

Thus, as successfully highlighted by SPIN, $\mathscr{L}(\mathscr{A}_{\mathscr{R}}^{\mathscr{MCS}}) \subseteq \mathscr{L}(\mathscr{A}_{\mathscr{R}}^{\mathscr{ACS}})$ does not hold for the VM design.

III. Furthermore, we used SPIN to prove that $\mathcal{L}(\mathcal{A}_{\mathcal{R}}^{\mathcal{ACS}}) \subseteq \mathcal{L}(\mathcal{A}_{\mathcal{R}}^{\mathcal{MCS}})$ also does not hold for the VM design. This implies that both implementations R_{MCS} and R_{ACS} of the VM design conform to its specification but one is not a refinement of the other. For some CAOS designs, such relationships are acceptable because *Correctness Requirement 3 (CR-3)* (Section 9.3) needs to be satisfied only if required based on the design requirements. For the VM design, R_{MCS} and R_{ACS} contain behaviors conforming to the specification and both implementations are acceptable.

These experiments demonstrate that the language containment-based verification approach of Algorithm *VerfLangCont* (Fig. 9.4) can be successfully used to compare behaviors of different implementations of CAOS designs.

9.6 Summary

Verification of hardware designs at a level of abstraction above RTL aids in faster and efficient verification early in the design cycle. In this chapter, we present an approach that provides such a verification path for CAOS designs. The proposed techniques can be used for verifying changes in the structure or behavior (such as those caused by the use of power minimization techniques proposed in the earlier chapters) of a CAOS design above RTL.

We propose the conversion of CAOS-based hardware designs into corresponding PROMELA models containing implementation-specific scheduling information. Such PROMELA models can then be verified using SPIN Model Checker for their essential properties. Moreover, for stronger language-containment proofs, we propose a technique that uses SPIN to verify if a particular implementation of the CAOS design is a refinement of its specification or some other implementation. We successfully used our verification techniques to check different CAOS designs for correctness and language containment-based proofs.

Note that for CAOS designs consisting of a large number of actions, the proposed SPIN-based verification techniques in this chapter might not scale well. However, a targeted model checker based on the presented verification techniques will scale better. In this chapter, our intent is to conceptually show how a model checker like SPIN can be used to verify CAOS-based hardware designs early in the design cycle.

Appendix: Algorithms and Code Listings

ALGORITHM: *GenVARS*. **IN:** CAOS Specification $\mathcal{S}$. **OUT:** Set of variables V.

---

1. Initialize $V = \phi$. (**Note:** V is the set of variables of PROMELA model $\mathcal{M}$.)
2. For each $s \in S$ such that $s \equiv (t, n, in)$,

   a. Construct a variable $v \equiv (t_p, n, in)$, where t_p is the PROMELA data-type corresponding to t.
   b. Construct a variable $v_{old} \equiv (t_p, n_{old}, in)$. (**Note:** During execution of model, v_{old} will be used to store old value of v for computations occurring in the bodies of processes.)
   c. Construct a variable $v_{old}^g \equiv (t_p, n_{old}^g, in)$. (**Note:** During execution of model, v_{old}^g will be used to store old value of v for computations of guards of processes.)
   d. Add v, v_{old} and v_{old}^g to V.

3. Add variables $v_{act} \equiv (byte, action, n+1)$ and $v_{act}^{Frd} \equiv (bool, action\_fired, False)$ to V, where n is the total number of actions $|A|$ of specification $\mathcal{S}$.
4. For each action $a \in A_m$ (**Note:** A_m is the set of interface methods.)

   a. Construct a variable $v \equiv (bool, m, False)$, where m is the name of action a. Add v to V.

5. Output V.

Fig. 9.5 Algorithm for generating set of variables of PROMELA model

---

ALGORITHM: *GenPROCS.* **INPUT:** 1.CAOS Specification $\mathscr{S}$, 2.Set of Variables V. **OUTPUT:** Set of processes P.

---

1. Initialize $P = \phi$. (**Note:** At the end, P will contain set of processes of PROMELA model $\mathscr{M}$.)
2. For each $a_i \in A$,

   a. Using variables of form $v_{old}^g \equiv (t_p, n_{old}^g, in)$ in set V, construct sets of variables V_g^i and V_b^i (with appropriate PROMELA data-types) corresponding to sets S_g^i and S_b^i of specification $\mathscr{S}$ respectively.
   b. Construct function $g_i^P = g_i(V_g^i)$ corresponding to guard of a_i.

3. For each $a_i \in A$,

   a. Initialize $B_i^P = \phi$, $B_i^{tmp} = \phi$.
   b. For each $s \in S_u^i$ such that $s \equiv (t, n, in)$
      i. Find variables $v \equiv (t_p, n, in)$ and $v_{old} \equiv (t_p, n_{old}, in)$ corresponding to s in set V.
      ii. Construct a PROMELA statement $stmt \equiv (v_{old}, v)$ denoting an assignment of the form $v_{old} = v$.
      iii. Add $stmt$ to B_i^P.
   c. For each $b \in B_i$ such that $b \equiv (s, b(S_b^i))$,
      i. Find a variable $v \equiv (t_p, n, in)$ corresponding to s in set V.
      ii. Construct a PROMELA statement $stmt \equiv (v, b(V_b^i))$ denoting an assignment of the form $v = b(V_b^i)$. (**Note:** $b(V_b^i)$ is constructed using types of PROMELA statements that retain corresponding meaning of $b(S_b^i)$.)
      iii. Add $stmt$ to B_i^{tmp}.
   d. Order statements of B_i^{tmp} such that they model the concurrent hardware behavior of action a_i. This may require using variables $v_{old} \in V$ which are used to store old values of variables corresponding to state elements of $\mathscr{S}$.
   e. $B_i^P = B_i^P \cup B_i^{tmp}$.
   f. Construct a PROMELA statement $stmt \equiv (\text{action\_fired}, \text{True})$ denoting an assignment of the form, action_fired = True.
   g. Add $stmt$ to B_i^P.
   h. If $a_i \in A_m$ then,
      i. $g_i^P = g_i^P$ (**Note:** For interface methods, no need to update guards with conflict information as their execution is decided by external module.)
   i. Else, for each $a_j \in C_i$,
      i. If $a_j \in A_m$ then,
         A. Find a corresponding variable $v \equiv (\text{bool}, m_j, in)$ in V, where m_j is name of action a_j.
         B. $g_i^P = g_i^P \,\&\&\, !v$.
      ii. Else, $g_i^P = g_i^P \,\&\&\, !g_j^P$.
   j. Add (m, B_i^P, g_i^P) to P, where $m = \text{IFC\_}m_i$ if $a_i \in A_m$, and $m = m_i$ otherwise (m_i is the name of action a_i).

4. Output P.

Fig. 9.6 Algorithm for generating set of processes of PROMELA model

---

ALGORITHM: *GenProcCycle.*
INPUT: 1. CAOS Specification $\mathscr{S}$, 2. Set of Variables V, 3. Set of processes P.
OUTPUT: PROMELA process *start_of_cycle.*

---

1. Initialize $B^P_{cycle} = \phi$.
2. For each variable $v \equiv (t_p, n, in)$ in set V,

   a. Find a variable $v^g_{old} \equiv (t_p, n^g_{old}, in)$ corresponding to v in V.
   b. Construct a PROMELA statement $stmt \equiv (v^g_{old}, v)$ denoting an assignment of the form $v^g_{old} = v$.
   c. Add $stmt$ to B^P_{cycle}.

3. For each $pr \in P$ such that $pr \equiv (m, B, g)$,

   a. **Comment:** Here, we generate statements corresponding to values read (inputs) from external modules of the hardware design. For verification purposes, all possible combinations of values are generated. Values are read at the start of every cycle.
   b. If m corresponds to the name of an action a such that $a \in A_m$ then,
      i. Find a variable $v \equiv (bool, m, in)$ in set V.
      ii. Construct a PROMELA statement $stmt_F \equiv (g, v, \text{False})$ denoting an assignment of the form $v = False$ if g computes to *False* or if some other higher priority process is enabled.
      iii. Construct a PROMELA statement $stmt_T \equiv (g, v, (\text{False, True}))$ denoting a nondeterministic assignment $v = False$ or $v = True$ if g computes to *True* and no other higher priority process is enabled.
      iv. Add $stmt_F$ and $stmt_T$ to B^P_{cycle}.

4. Using V, construct a set $STMT$ of PROMELA statements, if any, corresponding to resetting of any state elements of $\mathscr{S}$ at the start of a hardware clock cycle.
5. $B^P_{cycle} = B^P_{cycle} \cup STMT$.
6. Construct a PROMELA statement (action_fired, False) and add it to B^P_{cycle}.
7. Output (start_of_cycle, B^P_{cycle}, *True*).

Fig. 9.7 Algorithm for generating process denoting start of hardware cycle in PROMELA model

---

ALGORITHM: *AddSeqSched*.
INPUT: 1. CAOS Specification $\mathscr{S}$, 2. PROMELA Model $\mathscr{M}$ without scheduling information.
OUTPUT: PROMELA Model $\mathscr{M}_f$ executing processes based on *AOA Semantics*.

---

1. Initialize $P_s = \phi$.
2. Let V and P be the set of variables and processes of a PROMELA Model $\mathscr{M}$. Let $n = |P| - 1$.
3. For each $pr \in P$ such that $pr \equiv (m, B, g)$ and pr does not correspond to 'init' process.

   a. If m corresponds to 'start_of_cycle' then,
      i. Construct a PROMELA statement *cond* corresponding to comparison operation (action$==n$).
      ii. Construct a conditional expression $cond_d$ using set S of state elements of specification $\mathscr{S}$ such that when $cond_d$ is *True*, no action of $\mathscr{S}$ is enabled. If it is not possible to construct any such statement, go to Step (c).
      iii. Construct $cong_d^p$ corresponding to $cond_p$ in terms of variables in V.
      iv. Construct a PROMELA statement *stmt* performing assignment, action=n, if expression $cond_d^p$ evaluates to *True*, or assignment, action=0, otherwise.
      v. Add *stmt* to B.
   b. Else,
      i. If m corresponds to the name of an action a such that $a \in A_m$ then,
         A. Find a corresponding variable $v \equiv (bool, m, in)$ in V.
         B. Construct a PROMELA statement *cond* corresponding to comparison operation ((action$\neq n$) && v).
      ii. Else,
         A. Construct a PROMELA statement *cond* corresponding to comparison operation ((action$\neq n$) && g).
      iii. Construct a PROMELA statement *stmt* $\equiv$ (action, n) denoting an assignment action=n.
      iv. Add *stmt* to B.
   c. Construct a PROMELA atomic block, $Block_{atomic} \equiv atomic \{cond -> B\}$.
   d. Construct a PROMELA statement $Block_{do} \equiv \{do :: Block_{atomic}; od\}$ using repetition construct 'do'.
   e. Add $(m, Block_{do})$ to P_s.

4. Output PROMELA model $\mathscr{M}_f$ such that P_s is its set of processes and V is its set of variables.

Fig. 9.8 Algorithm for modeling AOA execution semantics in PROMELA model

---

ALGORITHM: *AddConcSched.*
INPUT: 1. CAOS Specification $\mathscr{S}$, 2. PROMELA Model $\mathscr{M}$ without scheduling information, 3. Sequential Ordering S_{order} of an implementation R, 4. Maximum number of actions max allowed to execute concurrently in R.
OUTPUT: PROMELA Model $\mathscr{M}_f$ based on scheduling information of R.

---

1. Initialize $P_s = \phi$, $i = 1$. Let V and P be the set of variables and processes of a PROMELA Model $\mathscr{M}$. Let $n = |P| - 1$.
2. Construct a variable $v \equiv$ (byte,num,0) (**Note:** *num* will be used to denote the number of processes executed so far in the model). Add v to V.
3. While $(i < n)$

   a. Find a process $pr \in P$ such that $pr \equiv (m,B,g)$ corresponds to i^{th} element of S_{order}.
      i. If m corresponds to the name of an action a such that $a \in A_m$ then,
         A. Find a corresponding variable $v \equiv$ (bool, m, in) in V.
         B. Construct a PROMELA statement $cond_B$ corresponding to comparison operation $((\text{action}== i)\ \&\&\ v)$.
      ii. Else,
         A. Construct a PROMELA statement $cond_B$ corresponding to comparison operation $((\text{action}== i)\ \&\&\ g)$.
      iii. Construct a PROMELA statement $stmt1 \equiv$ (num, num+1) denoting an assignment of the form, num=num+1.
      iv. Construct a PROMELA statement $stmt2 \equiv ((\text{num}== max), (\text{action}= n), (\text{action}= i + 1))$ denoting an assignment of the form action=n if (num==max) is *True*, and assignment of the form, action=i+1 otherwise. Add $stmt1$ and $stmt2$ to B.
      v. Construct a PROMELA statement block, $Block_B \equiv \{cond_B - > B\}$.
      vi. Construct a PROMELA statement $cond$ corresponding to comparison operation (action $== i$).
      vii. Construct a PROMELA statement $stmt3 \equiv$ (action, $i + 1$) denoting an assignment of the form, action = i+1.
      viii. Construct a PROMELA statement $stmt4 \equiv$ (action_fired, False) denoting an assignment of the form, action_fired = False. Let $B' = \{stmt3, stmt4\}$
      ix. Construct a PROMELA statement block, $Block_{B'} \equiv \{cond - > B'\}$.
      x. Construct a PROMELA atomic block, $Block_{atomic} \equiv atomic\ \{Block_{B'}\ unless\ Block_B\}$.
      xi. Construct a PROMELA statement $Block_{do} \equiv \{do :: Block_{atomic};\ od\}$ using repetition construct 'do'.
      xii. Add $(m, Block_{do})$ to P_s. $i = i + 1$.

4. For process $pr \in P$ such that $pr \equiv (m, B, True)$ corresponds to 'start_of_cycle' process,

   a. Construct a PROMELA statement $cond$ corresponding to comparison operation (action== n).
   b. Construct a PROMELA statement $stmt1 \equiv$ (action, 1) denoting an assignment of the form, action=1.
   c. Construct a PROMELA statement $stmt2 \equiv$ (num, 0) denoting an assignment of the form num=0. Add $stmt1$ and $stmt2$ to B.
   d. Construct a PROMELA atomic block, $Block_{atomic} \equiv atomic\ \{cond - > B\}$.
   e. Construct a PROMELA statement $Block_{do} \equiv \{do :: Block_{atomic};\ od\}$ using repetition construct 'do'. Add $(m, Block_{do})$ to P_s.

5. Output PROMELA model $\mathscr{M}_f$ such that P_s is its set of processes and V is its set of variables.

Fig. 9.9 Algorithm for modeling concurrent execution semantics in PROMELA model

Listing 9.1 PROMELA models for maximal and alternative concurrent schedules of VM design

```
int count , count_old ; byte action ;
bool moneyBack , gumControl , tenCentIn , fiftyCentIn , moneyBackButton ,
    action_fired , one_action_fired ;
proctype IFC_tenCentIn () {
  do
  :: atomic { { (action==1) -> {action = 2; action_fired = 0;} }
                unless
                { (action==1 && tenCentIn) ->
                    { count_old = count ;
                      count = count + 10;
                      one_action_fired = 1; action_fired = 1;
                      action =2; -FOR MAXIMAL CONCURRENT SCHEDULE
                      action =6; -FOR ALTERNATIVE CONC SCHEDULE
                    }
                } }
  od
}
proctype IFC_fiftyCentIn () {
  do
  :: atomic { { (action==2) -> {action = 3; action_fired = 0;} }
                unless
                { (action==2 && fiftyCentIn) ->
                    { count_old = count ;
                      count = count + 50;
                      one_action_fired = 1; action_fired = 1;
                      action =3; -FOR MAXIMAL CONCURRENT SCHEDULE
                      action =6; -FOR ALTERNATIVE CONC SCHEDULE
                    }
                } }
  od
}
proctype doDispenseMoney () {
  do
  :: atomic { { (action ==3) -> {action = 4; action_fired = 0;} }
                unless
                { (action==3 && moneyBack) ->
                    { count_old = count ;
                      if
                      :: (count == 0) -> moneyBack = 0;
                      :: else -> { count = count - 10;
                                   if
                                   :: (count == 0) ->
                                      moneyBack = 0;
                                   :: else -> skip ;
                                   fi }
                      fi
                      one_action_fired = 1; action_fired = 1;
                      action =4; -FOR MAXIMAL CONCURRENT SCHEDULE
                      action =6; -FOR ALTERNATIVE CONC SCHEDULE
                    }
                } }
  od
}
proctype doDispenseGum () {
  do
  :: atomic { { (action ==4) -> {action =5; action_fired = 0;} }
                unless
                { (action==4 && !tenCentIn && !fiftyCentIn
                  && !moneyBack && count >= 50) ->
                    { count_old = count ;
                      count = count - 50;
                      gumControl = 1;
                      one_action_fired = 1; action_fired = 1;
                      action =5; -FOR MAXIMAL CONCURRENT SCHEDULE
                      action =6; -FOR ALTERNATIVE CONC SCHEDULE
                    }
                } }
  od
}
```

Listing 9.2 PROMELA models for maximal and alternative concurrent schedules of VM design (continued)

```
proctype IFC_moneyBackButton() {
  do
  :: atomic { { (action==5) -> {action=6; action_fired = 0;} }
              unless
              { (action==5 && moneyBackButton) ->
                    { count_old = count;
                      moneyBack = 1;
                      one_action_fired = 1; action_fired = 1;
                      action = 6;
                    }
              } }
  od
}

proctype start_of_cycle() { /*Denotes end of a clock cycle.*/
  do
  :: atomic { if
              :: (action==6) ->
                    { if /* Read external stimulus */
                      :: moneyBack -> {tenCentIn=0; fiftyCentIn=0;
                            moneyBackButton=0;}
                      :: else -> if
                                 ::{ tenCentIn=1; fiftyCentIn=0;
                                       moneyBackButton=1;}
                                 ::{ tenCentIn=1; fiftyCentIn=0;
                                       moneyBackButton=0;}
                                 ::{ tenCentIn=0; fiftyCentIn=1;
                                       moneyBackButton=1;}
                                 ::{ tenCentIn=0; fiftyCentIn=1;
                                       moneyBackButton=0;}
                                 ::{ tenCentIn=0; fiftyCentIn=0;
                                       moneyBackButton=1;}
                                 ::{ tenCentIn=0; fiftyCentIn=0;
                                       moneyBackButton=0;}
                                 fi
                      fi
                      gumControl = 0;
                      one_action_fired = 0; action_fired = 0;
                      action = 1; }
              fi }
  od
}

init {
  atomic {
    count = 0; count_old = 0;
    moneyBack = 0; gumControl = 0;
    tenCentIn = 0; fiftyCentIn= 0; moneyBackButton = 0;
    action = 6;

    run IFC_tenCentIn(); run IFC_fiftyCentIn();
    run doDispenseMoney(); run doDispenseGum();
    run IFC_moneyBackButton(); run start_of_cycle(); }
}
```

Listing 9.3 LTL specification (in TLA style) encoding all behaviors of VM specification w.r.t. PROMELA models for maximal and alternative concurrent schedules

```
#define eqTo0  count == 0
#define gEq50  count >= 50
#define gt10   count > 10

#define inc10  count_old == (count - 10)
#define inc50  count_old == (count - 50)
#define dec10  count_old == (count + 10)
#define dec50  count_old == (count + 50)

#define ins10  tenCentIn
#define ins50  fiftyCentIn
#define mbbtn  moneyBackButton

#define dspGm  (dec50 && gumControl)


[] ( ( (moneyBack && gt10) -> X (!action_fired U (dec10 && moneyBack))
     ) &&
     ( (moneyBack && !gt10) -> X (!action_fired U (eqTo0 && !moneyBack))
     ) &&
     ( (( one_action_fired && !moneyBack && gEq50) -> X (!action_fired U
            (inc10 || inc50 || moneyBack || dspGm))) &&
       ((!one_action_fired && !moneyBack && gEq50 && !ins10 && !ins50 &&
            mbbtn) -> X (!action_fired U (moneyBack || dspGm))) &&
       ((!one_action_fired && !moneyBack && gEq50 && !ins10 && !ins50 &&
            !mbbtn) -> X (!action_fired U dspGm))
     ) &&
     ( (( one_action_fired && !moneyBack && !gEq50) -> X (!action_fired
            U (inc10 || inc50 || moneyBack))) &&
       ((!one_action_fired && !moneyBack && !gEq50 && !ins10 && !ins50
            && mbbtn) -> X (!action_fired U moneyBack)) &&
       ((!one_action_fired && !moneyBack && !gEq50 && !ins10 && !ins50
            && !mbbtn) -> X (!action_fired U (inc10 || inc50 ||
            moneyBack)))
     ) &&
     ( ((!one_action_fired && !moneyBack && ins10 && !ins50 && mbbtn)
            -> X (!action_fired U (inc10 || moneyBack))) &&
       ((!one_action_fired && !moneyBack && ins10 && !ins50 && !mbbtn)
            -> X (!action_fired U inc10)) &&

       ((!one_action_fired && !moneyBack && !ins10 && ins50 && mbbtn)
            -> X (!action_fired U (inc50 || moneyBack))) &&
       ((!one_action_fired && !moneyBack && !ins10 && ins50 && !mbbtn)
            -> X (!action_fired U inc50))
     )
  )
```

Chapter 10
Epilogue

In the past, high-level synthesis from CAOS has been shown to produce designs optimized for area and latency. However, not much work had been done in the area of synthesis of low-power hardware designs from CAOS and their verification. This book focuses on solving the problems of generation of power-optimized hardware using CAOS and verification of the synthesized low-power hardware.

Power consumption of hardware designs has become a critical metric that should be taken into consideration while evaluating the viability and success of any synthesis process. Although, traditional RTL-to-gate synthesis tools can be used for synthesizing low-power hardware, as design complexities increase, using high-level (above RTL) models to solve the low-power optimization problem at a higher level of abstraction provides a more efficient solution. In the first part of this book, we target the minimization of dynamic power as well as peak power consumption of hardware designs generated using CAOS.

For solving the low-power optimization problems related to CAOS-based synthesis, we formalized these problems, analyzed their complexities, and proposed heuristics to efficiently solve these problems at a higher (above RTL) level of abstraction. Scheduling is the most important phase for any high-level synthesis tool which can affect the power consumption as well as latency of a hardware design. For this reason, we also presented a detailed complexity analysis of scheduling problem related to CAOS-based synthesis under the given power and latency constraints and proposed efficient heuristics to solve it.

Its known that, at higher abstraction levels, opportunities for design space exploration are much more because as we descend the abstraction levels, more design decisions get fixed. Therefore, the power optimization opportunities increase as we go up in the abstraction levels. The caveat is that power consumption depends a lot on lower level design parameters, such as technology, cell libraries, device characteristics, layout. This makes it harder to choose between these different alternatives, even though there are more such alternatives available at high level. Assuming that we have a concrete idea about some of the implementation parameters and ways to estimate power consumption by various alternatives at higher abstraction levels, such power optimizing synthesis opens up a lot of possibilities.

Once we have solved such high-level estimation problems (great progress has been made on this topic both in industry and academia in the past few years),

G. Singh, S.K. Shukla, *Low Power Hardware Synthesis from Concurrent Action-Oriented Specifications*, DOI 10.1007/978-1-4419-6481-6_10,
© Springer Science+Business Media, LLC 2010

one could use the techniques proposed in this book for faster architectural exploration during synthesis. Since Bluespec language and semantics are abstracted into our CAOS model, we implemented some of our proposed low-power heuristics in the *Bluespec Compiler* (*BSC*) which is a commercial high-level synthesis tool. We tested the power savings obtained by those heuristics on some realistic CAOS-based hardware designs and present experimental results along with detailed analysis in this book.

Since low-power optimizations usually involve change in the structure and/or behavior of synthesized hardware designs, verification of power-optimized hardware designs become imperative. For this, in the later part of this book, we solve the problem of formal verification of hardware designs generated using CAOS. The changes made to the structure/behavior of a hardware design, as part of the power-reduction algorithms proposed in this book, can be effectively verified using the SPIN Model Checker as demonstrated in Chapter 9. The proposed technique enables fast and early formal verification of CAOS-based hardware designs at a level of abstraction above RTL.

The work described in this book also paves the way for future research in the area of generation of power-optimized hardware using CAOS-based high-level synthesis and formal verification of such designs. This work can be extended in the following directions

1. In literature, lots of techniques have been proposed in order to minimize the dynamic power consumption of hardware designs. Apart from the dynamic power reduction techniques proposed in this book, some other techniques like multiple voltage domains, sequential clock-gating can be investigated to further reduce the dynamic power component of CAOS-based designs.
2. Peak power reductions achieved using the heuristic proposed in Chapter 8 are associated with a corresponding increase in the latency of various designs. In that work, only peak power constraint of a design is used to schedule various actions of the design accordingly. As an extension of that work, the peak power optimization problem can be formulated as a bi-criteria optimization problem using the peak power constraint as well as the latency constraint of the design to schedule its actions. Such formulation can then be used to find various Pareto points based on the peak power and latency requirements of the design.
3. With the shrinking geometries, these days leakage power is starting to be a dominant component of the total power consumption of hardware designs. Power-gating is one of the most common technique used for targeting the reduction of leakage power of a design. The low-power work presented in this book can be extended to solve the leakage power optimization problem for CAOS-based designs at a level of abstraction above RTL.
4. Accurate power estimation at the high abstraction level of CAOS will greatly aid in taking efficient power minimizing decisions during CAOS-based synthesis. This is another research area having the potential of gaining a lot of attention in future.

5. Other techniques for formal verification of CAOS-based hardware designs can also be investigated. Since CAOS-based synthesis is relatively new and is gaining momentum in the hardware industry, a dedicated tool targeting the verification of such hardware designs can be developed.

References

1. S. Ahuja, S. T. Gurumani, C. Spackman, and S. K. Shukla. Hardware Coprocessor Synthesis from an ANSI C Specification. *IEEE Design and Test of Computers*, 26:58–67, 2009.
2. S. Ahuja, D. A. Mathaikutty, A. Lakshminarayana, and S. Shukla. Statistical Regression Based Power Models for Co-processors for Faster and Accurate Power Estimation. In *Proceedings of the 22nd IEEE International SOC Conference*, pp. 399–402, 2009.
3. S. Ahuja, D. A. Mathaikutty, and S. Shukla. Applying Verification Collaterals for Accurate Power Estimation. In *Proceedings of the 9th International Workshop on Microprocessor Test and Verification (MTV)*, Austin, TX, USA, pp. 61–66, 2008.
4. S. Ahuja, D. A. Mathaikutty, S. Shukla, and A. Dingankar. Assertion-Based Modal Power Estimation. In *Proceedings of the 8th International Workshop on Microprocessor Test and Verification (MTV)*, Austin, TX, USA, pp. 3–7, 2007.
5. S. Ahuja, D. A. Mathaikutty, and S. K. Shukla. SCoPE: Statistical Regression Based Power Models for Co-Processors Power Estimation. *Journal of Low Power Electronics*, 5(4): 407–415, December 2009.
6. S. Ahuja, D. A. Mathaikutty, G. Singh, J. Stetzer, S. Shukla, and A. Dingankar. Power Estimation Methodology for a High-Level Synthesis Framework. In *Proceedings of the 10th International Symposium on Quality Electronics Design (ISQED)*, San Jose, CA, USA, pp. 541–546, 2009.
7. S. Ahuja and S. K. Shukla. MCBCG: Model Checking Based Sequential Clock-Gating. In *IEEE International High Level Design Validation and Test Workshop (HLDVT)*, San Francisco, CA, USA, pp. 20–25, November 2009.
8. S. Ahuja, W. Zhang, and S. K. Shukla. System Level Simulation Guided Approach to Improve the Efficacy of Clock-Gating. *FERMAT Technical Report*, Virginia Tech, Blacksburg, VA, USA.
9. S. Ahuja, W. Zhang, and S. K. Shukla. A Methodology for Power Aware High-Level Synthesis of Co-processors from Software Algorithms. In *Proceedings of International VLSI Design Conference*, Bangalore, India, 2010, pp. 282–287, January 2010.
10. A. Arvind, R. Nikhil, D. Rosenband, and N. Dave. High-Level Synthesis: An Essential Ingredient for Designing Complex ASICs. In *Proceedings of the International Conference on Computer Aided Design (ICCAD'04)*, San Jose, CA, USA, pp. 775–782, November 2004.
11. F. Baader and T. Nipkow. *Term Rewriting and All That*. Cambridge University Press, Cambridge, 1998.
12. B. S. Baker. Approximation Algorithms for NP-Complete Problems on Planar Graphs. *Journal of the Association for Computing Machinery*, 41:153–180, 1994.
13. P. Berman and T. Fujito. Approximating Independent Sets in Degree 3 Graphs. *Proceedings of the 4th Workshop on Algorithms and Data Structures, Lecture Notes in Computer Science*, 955:449–460, 1995.
14. G. Berry. Esterel on Hardware. *Philosophical Transactions of the Royal Society of London (Series A)*, 339:87–104, April 1992.

15. G. Berry and G. Gonthier. The Esterel Synchronous Programming Language: Design, Semantics, Implementation. *Science of Computer Programming*, 19(2):87–152, November 1992.

16. J. Bingham, J. Erickson, G. Singh, and F. Andersen. Industrial Strength Refinement Checking. *Formal Methods in Computer Aided Design (FMCAD)*, Austin, TX, USA, November 2009.

17. J. Blazewicz, J. K. Lenstra, and A. H. G. Rinooy Kan. Scheduling Subject to Resource Constraints: Classification and Complexity. *Discrete Applied Mathematics*, 5:11–24, 1983.

18. Bluespec Inc. http://www.bluespec.com/. BluespecCompiler.

19. J. Brandt, K. Schneider, S. Ahuja, and S K. Shukla. The Model Checking View to Clock Gating and Operand Isolation. In *Accepted at the International Conference on Application of Concurrency to System Design*, June 2010.

20. R.E. Bryant. Graph-Based Algorithms for Boolean Function Manipulation. *IEEE Transactions on Computers,* 35(8):677–691, 1986.

21. G. Campers, O. Henkes, and P. Leclerq. Graph Coloring Heuristics: A Survey, Some New Propositions and Computational Experiences on Random and Leighton's Graphs. *Proceedings of the Operational Research, 87, Buenos Aires*, pp. 917–932, 1987.

22. Cebatech Inc. C2R Compiler. http://www.cebatech.com

23. Celoxica Limited. http://www.celoxica.com. Agility Compiler – Advanced Synthesis Technology For SystemC.

24. Celoxica. Handel-C Language Reference Manual RM-1003-4.0, 2003. http://www.celoxica.com

25. S. Chaki, E. Clarke, A. Groce, J. Ouaknine, O. Strichman, and K. Yorav. Efficient Verification of Sequential and Concurrent C Programs. *Formal Methods in System Design,* 25(2/3): 129–166, 2004.

26. S. Chaki, E. M. Clarke, A. Groce, S. Jha, and H. Veith. Modular Verification of Software Components in C. *IEEE Transactions on Software Engineering,* 30(6):388–402, June 2004.

27. A. P. Chandrakasan, M. Potkonjak, R. Mehra, J. Rabaey, and R. W. Brodersen. Optimizing Power Using Transformations. *IEEE Transactions on Computer-Aided Design,* 14:12–31, January 1995.

28. J.-M. Chang and M. Pedram. *Power Optimization and Synthesis at Behavioral and System Levels Using Formal Methods.* Kluwer Academic Publishers, Norwell, MA, 1999.

29. A. Chattopadhyay, B. Geukes, D. Kammler, E. M. Witte, O. Schliebusch, H. Ishebabi, R. Leupers, G. Ascheid, and H. Meyr. Automatic ADL-Based Operand Isolation for Embedded Processors. *Design Automation and Test in Europe (DATE'06)*, Germany 2006.

30. D. Chen, J. Cong, and Y. Fan. Low-Power High-Level Synthesis for FPGA Architectures. In *IEEE/ACM International Conference on Low Power Electronics and Design (ISLPED'03)*, Seoul, Korea, pp. 134–139, August 2003.

31. T.-H. Chiang, L.-R. Dung, and M.-F. Yaung. Modeling and Formal Verification of Dataflow Graph in System-Level Design Using Petri Net. In *IEEE International Symposium on Circuits and Systems (ISCAS'05)*, vol. 6, Kobe, Japan, pp. 5674–5677, May 2005.

32. E. Clarke, O. Grumberg, M. Talupur, and D. Wang. High Level Verification of Control Intensive Systems Using Predicate Abstraction. In *First ACM and IEEE International Conference on Formal Methods and Models for Co-design (MEMOCODE'03)*, France, pp. 55–64, June 2003.

33. E. Clarke and D. Kroening. Hardware Verification Using ANSI-C Programs as a Reference. In *Asia and South Pacific Design Automation Conference (ASP-DAC'03)*, Kitakyushu, Japan, pp. 308–311, January 2003.

34. E. Clarke, D. Kroening, and K. Yorav. Behavioral Consistency of C and Verilog Programs Using Bounded Model Checking. In *Design Automation Conference (DAC'03)*, Anaheim, CA, USA, pp. 368–371, June 2003.

35. E. M. Clarke and E. A. Emerson. Design and Synthesis of Synchronization Skeletons Using Branching Time Temporal Logic. In *IBM Workshop on Logics of Programs*, vol. 131 of LNCS, pp. 52–71. Springer-Verlag, Heidelberg, 1981.

36. E. M. Clarke, E. A. Emerson, and A. P. Sistla. Automatic Verification of Finite-State Concurrent Systems Using Temporal Logic Specifications. In *ACM Transactions on Programming Languages and Systems*, vol. 8, pp. 244–263, 1986.
37. E. M. Clarke, O. Grumberg, and D. A. Peled. *Model Checking*. The MIT Press, Cambridge, MA, 2000.
38. E. M. Clarke, A. Biere, R. Raimi, and Y. Zhu. Bounded Model Checking Using Satisfiability Solving. *Formal Methods in System Design*, 19(1):7–34, 2001.
39. E. G. Coffman, Jr., M. R. Garey, and D. S. Johnson. Approximation Algorithms for Bin-Packing – A Survey. In: D. S. Hochbaum (ed.) *Approximation Algorithms for NP-hard Problems*. PWS Publishing Company, Boston, MA, pp. 46–93, 1997.
40. B. Cook, D. Kroening, and N. Sharygina. Accurate Theorem Proving for Program Verification. In *ETH Technical Report 464, ETH Zurich, Department of Computer Science*, Zurich, Switzerland, 2006.
41. T. Cormen, C. E. Leiserson, R. Rivest, and C. Stein. *Introduction to Algorithms, 2nd edn.* MIT Press and McGraw-Hill, Cambridge, MA, 2001.
42. O. Coudert, J. C. Madre, and C. Berthet. Verifying Temporal Properties of Sequential Machines Without Building Their State Diagrams. In *Proceedings of the 10th International Computer Aided Verification Conference*, New Brunswick, NJ, USA, pp. 23–32, 1990.
43. D. Dal, D. Kutagulla, A. Nunez, and N. Mansouri. Power Islands: A High-Level Synthesis Technique for Reducing Spurious Switching Activity and Leakage. In *Proceedings of the 48th Midwest Symposium on Circuits and Systems, 2005*, vol 2, pp. 1875–1879, August 2005.
44. D. L. Dill, A. J. Drexler, A. J. Hu, and C. H. Yang. Protocol Verification as a Hardware Design Aid. In *IEEE International Conference on Computer Design: VLSI in Computers and Processors*, pp. 522–525. IEEE Computer Society, Washington, DC, 1992.
45. G. De Micheli. Hardware Synthesis from C/C++ Models. In *Design Automation and Test in Europe Conference and Exhibition 1999 (DATE'99)*, Munich, Germany, pp. 382–383, March 1999.
46. G. De Micheli, D. Ku, F. Mailhot, and T. Truong. The Olympus Synthesis System. *IEEE Design and Test of Computers*, 7:37–53, October 1990.
47. R. Drechsler and D. Grosse. Reachability Analysis for Formal Verification of System C. In *Proceedings of the Euromicro Symposium on Digital System Design (DSD'02)*, Germany, pp. 337–340, September 2002.
48. S. A. Edwards. High-Level Synthesis from the Synchronous Language Esterel. In *Proceedings of the International Workshop of Logic and Synthesis, New Orleans, Louisiana (IWLS'02)*, pp. 401–406, June 2002.
49. S. A. Edwards. The Challenges of Hardware Synthesis from C-Like Languages. In *Proceedings of the International Workshop on Logic Synthesis, Temecula, CA (IWLS'04)*, pp. 509–516, June 2004.
50. Esterel Technologies. http://www.esterel-eda.com/. Esterel Studio.
51. U. Feige and J. Kilian. Zero Knowledge and the Chromatic Number. *Journal of Computer and System Sciences*, 57:187–199, 1998.
52. Forte Design Systems. http://www.forteds.com. Cynthesizer.
53. D. D. Gajski, J. Zhu, R. Dömer, A. Gerstlauer, and S. Zhao. *SpecC: Specification Language and Methodology*. Kluwer Publications, Dordrecht, 2000.
54. M. R. Garey and D. S. Johnson. *Computers and Intractability: A Guide to the Theory of NP-Completeness*. W. H. Freeman and Company, San Francisco, CA, 1979.
55. GEZEL Reference Manual. GEZEL Language Information. http://rijndael.ece.vt.edu/gezel2/index.php/Main_Page.
56. T. Grotker, S. Liao, G. Martin, and S. Swan. *System Design with SystemC*. Kluwer Publications, Boston, MA, 2002.
57. S. Gupta, R. K. Gupta, N. D. Dutt, and A. Nicolau. *SPARK: A Parallelizing Approach to the High-Level Synthesis of Digital Circuits*. Kluwer Academic Publisher, Dordrecht, 2004.

58. L. A. Hall. Approximation Algorithms for Scheduling. In: D. S. Hochbaum (ed.) *Approximation Algorithms for NP-Hard Problems*. PWS Publishing Company, Boston, MA, pp. 1–45, 1997.

59. M. M. Halldorsson. Approximations of Weighted Independent Set and Hereditary Subset Problems. In *Proceedings of the 5th Annual International Conference on Computing and Combinatorics, LNCS*, pp. 261–270. Springer-Verlag, Heidelberg, Germany, 1999.

60. J. Hastad. Clique is Hard to Approximate Within $n^{1-\varepsilon}$. *Acta Mathematica*, 182:105–142, 1999.

61. A. Hertz and D. de Werra. Using Tabu Search Techniques for Graph Coloring. *Computing*, 39:345–351, 1987.

62. J. C. Hoe and A. Arvind. Hardware Synthesis from Term Rewriting Systems. In *Proceeding of VLSI'99 Lisbon, Portugal*, December 1999.

63. G. J. Holzmann. The Model Checker SPIN. *Software Engineering*, 23(5):279–295, 1997.

64. G. J. Holzmann. *The SPIN Model Checker*. Addison Wesley, Boston, MA, 2004.

65. H. B. Hunt, III, M. V. Marathe, V. Radhakrishnan, S. S. Ravi, D. J. Rosenkrantz, and R. E. Stearns. NC-Approximation Schemes for NP- and PSPACE-Hard Problems for Geometric Graphs. *Journal of Algorithms*, 26(2):238–274, February 1998.

66. H. B. Hunt, III, M. V. Marathe, V. Radhakrishnan, S. S. Ravi, D. J. Rosenkrantz, and R. E. Stearns. Parallel Approximation Schemes for a Class of Planar and Near Planar Combinatorial Problems. *Information and Computation*, 173(1):40–63, February 2002.

67. J. Ivers and N. Sharygina. Overview of ComFoRT: A Model Checking Reasoning Framework. In *CMU/SEI-2004-TN-018*, 2004.

68. H. Jain, D. Kroening, and E. Clarke. Verification of SpecC Using Predicate Abstraction. In *Second ACM and IEEE International Conference on Formal Methods and Models for Co-design (MEMOCODE'04)*, San Diego, CA, USA, pp. 7–16, June 2004.

69. J. Monteiro, S. Devadas, P. Ashar, and A. Mauskar. Scheduling Techniques to Enable Power Management. In *Proceedings of Design Automatin Conference*, Las Vegas, NV, USA, pp. 349–352, June 1996.

70. K. S. Khouri, G. Lakshminarayana, and N. K. Jha. High-Level Synthesis of Low-Power Control-Flow Intensive Circuits. *IEEE Transactions on Computer-Aided Design (TCAD'99)*, 18(12): 1715–1729, December 1999.

71. D. Kroening and E. Clarke. Checking Consistency of C and Verilog Using Predicate Abstraction and Induction. In *Proceedings of the 2004 IEEE/ACM International Conference on Computer Aided Design (ICCAD'04)*, San Jose, CA, USA, pp. 66–72, November 2004.

72. D. Kroening, A. Groce, and E. M. Clarke. Counterexample Guided Abstraction Refinement via Program Execution. In *Proceedings of the 6th International Conference on Formal Engineering Methods*, Seattle, WA, USA, pp. 224–238, November 2004.

73. D. Kroening and N. Sharygina. Formal Verification of SystemC by Automatic Hardware/-Software Partitioning. In *Proceedings of Third ACM and IEEE International Conference on Formal Methods and Models for Co-design (MEMOCODE'05)*, Verona, Italy, pp. 101–110, July 2005.

74. D. Ku and G. De Micheli. Hardware C – A Language for Hardware Design (Version 2.0). *Technical Report: CSL-TR-90-419*, Stanford University, Stanford, CA. August 1990.

75. A. Kumar and M. Bayoumi. Multiple Voltage-Based Scheduling Methodology for Low-Power in the High-Level Synthesis. *IEEE International Symposium on Circuits and Systems*, Orlando, FL, USA, 1:371–374, June 1999.

76. R. Kurki-Suonio. *A Practical Theory of Reactive Systems: Incremental Modeling of Dynamic Behaviors*. Springer, Secaucus, NJ, 1998.

77. A. Lakshminarayana, S. Ahuja, and S. Shukla. Coprocessor Design Space Exploration Using High Level Synthesis. In *Proceedings of the 10th International Symposium on Quality Electronics Design (ISQED)*, San Jose, CA, USA, 2010.

78. G. Lakshminarayana, A. Raghunathan, N. K. Jha, and S. Dey. A Power Management Methodology for High-Level Synthesis. In *Proceedings of the 11th International Conference on VLSI Design*, Chennai, India, pp. 24–29, January 1998.

79. L. Lamport. The Temporal Logic of Actions. *ACM Transactions on Programming Languages and Systems*, 16(3):872–923, May 1994.
80. M. Liz et al. Efficient Generation of Schedulers for Guarded Atomic Actions. *Technical Memo, Bluespec Inc.*, 2005. http://www.bluespec.com/
81. K. L. Man. Enhancing Formal Methods for SystemC Designs. In *Proceedings of the 4th PROGRESS Symposium on Embedded Systems (Nieuwegein, The Netherlands)*, pp. 141–146, October 2003.
82. Mentor Graphics. http://www.mentor.com/. Catapult C Synthesis.
83. J. Misra. *A Discipline of Multi-Programming*. Springer, New York, 2001.
84. S. P. Mohanty and N. Ranganathan. A Framework for Energy and Transient Power Reduction During Behavioral Synthesis. In *Proceedings of the International Conference on VLSI Design*, New Delhi, India, pp. 539–545, 2003.
85. S P. Mohanty, N. Ranganathan, and S. K. Chappidi. ILP Models for Energy and Transient Power Minimization During Behavioral Synthesis. In *Proceedings of the International Conference on VLSI Design*, Mumbai, India, p. 745, 2004.
86. M. Munch, B. Wurth, R. Mehra, J. Sproch, and N. Wehn. Automating RT-Level Operand Isolation to Minimize Power Consumption in Datapaths. In *Design Automation and Test in Europe (DATE'00)*, Paris, France, 2000.
87. A. K. Murugavel and N. Ranganathan. A Game-Theoretic Approach for Binding in Behavioral Synthesis. In *Proceedings of the 16th International Conference on VLSI Design*, New Delhi, India, pp. 452–458, January 2003.
88. R. Dömer, A. Gerstlauer, and D. Gajski. SpecC Language Reference Manual, Version 2.0. March 2001.
89. M. Pedram and A. Abdollahi. Low Power RT-Level Synthesis Techniques – A Tutorial, Department of Electrical Engineering, University of Southern California, May 2005.
90. C. Pixley. A Computational Theory and Implementation of Sequential Hardware Equivalence. In *Proceedings of the 1990 CAV Conference, DIMACS Series on Discrete Mathematics and Theoretical Computer Science, American Mathematical Society*, New Brunswick, NJ, USA, pp. 293–320, 1990.
91. P. Prabhakaran and P. Banerjee. Parallel Algorithms for Simultaneous Scheduling, Binding and Floorplanning in High-Level Synthesis. In *IEEE International Symposium on Circuits and Systems (ISCAS'98)*, vol. 6, Monterey, CA, USA, pp. 372–376, May 1998.
92. J. P. Quielle and J. Sifakis. Specification and Verification of Concurrent Systems in CESAR. In *Proceedings of the 5th International Symposium on Programming, Lecture Notes In Computer Science*, vol. 137, pp. 337–351, 1982.
93. A. Raghunathan and N. K. Jha. Behavioral Synthesis for Low Power. In *Proceedings of the 1994 IEEE International Conference on Computer Design: VLSI in Computer & Processors (ICCAD'94)*, San Jose, CA, USA, pp. 318–322, 1994.
94. A. Raghunathan, N. K. Jha, and S. Dey. Background. In *High-Level Power Analysis And Optimization*, pp. 18–21. Kluwer Academic Publishers, Norwell, MA, 1998.
95. A. Raghunathan, N. K. Jha, and S. Dey. *High-Level Power Analysis and Optimization*. Kluwer Academic Publishers, Norwell, MA, 1998.
96. V. Raghunathan, S. Ravi, A. Raghunathan, and G. Lakshminarayana. Transient Power Management Through High Level Synthesis. In *Proceedings of the ICCAD*, San Jose, CA, USA, pp. 545–552, 2001.
97. D. Rosenband and A. Arvind. Modular Scheduling of Guarded Atomic Actions. In *Proceedings of the Design Automation Conference (DAC'04)*, San Diego, CA, USA, June 2004.
98. T. Sakunkonchak, S. Komatsu, and M. Fujita. Verification of Synchronization in SpecC Description with the Use of Difference Decision Diagrams. *IEICE Transactions on Fundamentals of Electronics, Communications and Computer Sciences*, E86-A(12):3192–3199, December 2003.
99. Sequence Design Inc. http://www.sequencedesign.com/
100. W. T. Shiue. High Level Synthesis for Peak Power Minimization Using Ilp. In *Proceedings of the IEEE International Conference on ASSAP*, Boston, MA, USA, pp. 103–112, 2000.

101. W.-T. Shiue and C. Chakrabarti. ILP-Based Scheme for Low Power Scheduling and Resource Binding. In *IEEE International Symposium on Circuits and Systems, Geneva, Switzerland (ISCAS'00)*, vol. 3, Geneva, Switzerland, pp. 279–282, May 2000.

102. W.-T. Shiue and C. Chakrabarti. Low-Power Scheduling with Resources Operating at Multiple Voltages. *IEEE Transactions On Circuits and Systems – II: Analog and Digital Signal Processing*, 47(6):536–543, June 2000.

103. G. Singh. Optimization and Verification Techniques for Hardware Synthesis from Concurrent Action-Oriented Specifications, Ph.D Dissertation, Virginia Tech, Blacksburg, VA, USA, 2008.

104. G. Singh, S. Gupta, S. K. Shukla, and R. Gupta. Parallelizing High-Level Synthesis: A Code Transformational Approach to High-Level Synthesis. *Chapter 11 of CRC Handbook on EDA for IC System Design, Verification and Testing*. CRC Press, Taylor & Francis Group, 2006.

105. G. Singh, J. B. Schwartz, S. Ahuja, and S. K. Shukla. Techniques for Power-Wware Hardware Synthesis from Concurrent Action Oriented Specifications. *Journal of Low Power Electronics (JOLPE)*, 3(2):156–166, August 2007.

106. G. Singh, J. B. Schwartz, and S. K. Shukla. A Formally Verified Peak-Power Reduction Technique for Hardware Synthesis from Concurrent Action-Oriented Specifications. *Journal of Low Power Electronics (JOLPE)*, 5(2):135–144, August 2009.

107. G. Singh and S. K. Shukla. Low-Power Hardware Synthesis from TRS-Based Specifications. In *Fourth ACM and IEEE International Conference on Formal Methods and Models for Codesign (MEMOCODE'06), Napa Valley, CA, USA*, pp. 49–58, July 2006.

108. G. Singh and S. K. Shukla. Algorithms for Low Power Hardware Synthesis from CAOS – Concurrent Action Oriented Specifications. *Special Issue of International Journal of Embedded Systems on Power/Energy/Thermal topics (IJES'07)*, 3(1–2):83–92, 2007.

109. G. Singh and S. K. Shukla. Model Checking Bluespec Specified Hardware Designs. In *Proceedings of the 8th International Workshop on Microprocessor Test and Verification (MTV), Austin, TX, USA*, December 2007.

110. G. Singh and S. K. Shukla. Verifying Compiler Based Refinement of Bluespec Specifications using the SPIN Model Checker. In *Proceedings of the 15th International SPIN Workshop on Model Checking of Software (SPIN), Los Angeles, CA, USA*, pp. 250–269, August 2008.

111. G. Singh, S. S. Ravi, S. Ahuja, and S. K. Shukla. Complexity of Scheduling in Synthesizing Hardware from Concurrent Action Oriented Specifications. *Power-Aware Computing Systems, Dagstuhl Seminar Proceedings 07041, Dagstuhl, Germany*, 2007.

112. SMV. http://www-cad.eecs.berkeley.edu/~kenmcmil/.

113. C. E. Stroud, R. R. Munoz, and D. A. Pierce. Behavioral Model Synthesis with Cones. *IEEE Design and Test of Computers*, 5:22–30, June 1988.

114. Synfora. http://www.synfora.com/. PICO Express.

115. J. Uchida, N. Togawa, M. Yanagisawa, and T. Ohtsuki. A Thread Partitioning Algorithm in Low Power High-Level Synthesis. In *Asia and South Pacific Design Automation Conference (ASP-DAC'04), Yokohama, Japan*, pp. 74–79, January 2004.

116. J. D. Ullman. NP-Complete Scheduling Problems. *Journal of Computer and System Sciences*, 10(3):384–393, June 1975.

117. K. Wakabayashi. C-Based Synthesis Experiences with a Behavior Synthesizer, "Cyber". In *Design Automation and Test in Europe Conference and Exhibition 1999 (DATE'99), Munich, Germany*, pp. 390–393, March 1999.

118. D. B. West. *Introduction to Graph Theory*. 2nd edn.. Prentice Hall, Inc., Englewood Cliffs, NJ, 2001.

119. Xilinx Inc. Spartan-3E Starter Kit. http://www.xilinx.com/products/devkits/HW-SPAR3E-SK-US-G.htm

120. L. Zhong and N. K. Jha. Interconnect-Aware High-Level Synthesis for Low Power. In *IEEE/ACM International Conference on Computer Aided Design (ICCAD'02), San Jose, CA, USA*, pp. 110–117, November 2002.

121. D. Zuckerman. Linear Degree Extractors and the Inapproximability of Max Clique and Chromatic Number. In *Proceedings of the ACM International Symposium on Theory of Computing (STOC'00), Seattle, WA, USA*, pp. 681–690, May 2006.

Index